中外建筑与城市文化系列丛书

文化视角下的城市空间
URBAN SPACE IN CULTURAL PERSPECTIVE

刘健 刘佳燕 等 著

清华大学出版社
北京

版权所有,侵权必究。举报:010-62782989,beiqinquan@tup.tsinghua.edu.cn。

图书在版编目(CIP)数据

文化视角下的城市空间:汉英对照 / 刘健等著. —北京:清华大学出版社,2021.1(2021.12重印)
(中外建筑与城市文化系列丛书)
ISBN 978-7-302-56730-1

Ⅰ.①文… Ⅱ.①刘… Ⅲ.①城市空间-研究-汉、英 Ⅳ.①TU984.11

中国版本图书馆CIP数据核字(2020)第210744号

责任编辑:张占奎
装帧设计:陈国熙
责任校对:赵丽敏
责任印制:丛怀宇

出版发行:清华大学出版社
网　址:http://www.tup.com.cn, http://www.wqbook.com
地　址:北京清华大学学研大厦A座
邮　编:100084
社 总 机:010-62770175
邮　购:010-62786544
投稿与读者服务:010-62776969, c-service@tup.tsinghua.edu.cn
质量反馈:010-62772015, zhiliang@tup.tsinghua.edu.cn
印 装 者:北京博海升彩色印刷有限公司
经　销:全国新华书店
开　本:285mm×210mm
印　张:11.25
字　数:423千字
版　次:2021年3月第1版
印　次:2021年12月第2次印刷
定　价:88.00元

产品编号:057201-01

小小课堂　大千世界
——"中外建筑与城市文化系列丛书"序

A SMALL CLASS MINIATURING A BIG WORLD
PREFACE TO A SERIES ON ARCHITECTURAL AND URBAN CULTURE OF EAST AND WEST

朱文一
教授，学术委员会主任
清华大学建筑学院

Zhu Wenyi
Prof.
Director, Academic Committee
School of Architecture, Tsinghua University

时间：2020 年 2 月 10 日
Time：Feb. 10, 2020

这是一门非常专业的课程，不仅是专业领域之专业，而且还表现在其全英文授课之专业上。选课学生以清华大学建筑学院英文授课建筑学硕士班（EPMA）学生为主。然而，几十名学生组成的小小课堂，却容纳了建筑学、城乡规划学、风景园林学等学科构成的大千世界。更加难得的是，授课教师多是来自世界各地的、在各自学科领域卓有成就的大教授、大专家们。

这就是由清华大学建筑学院副院长刘健教授主持的课程"中外建筑与城市文化比较"。

印象中，刘健老师做事非常严谨、认真。提前谋划、细致筹划和长远计划是她一贯的做事风格。通过多年点滴积淀，积沙成塔，将一门课程的内容变成一本内容丰富的专业书。这本书的出版再一次体现了刘老师的行事风格。在此，为她点赞！

说起来，我和这本书还有一点关联。首先，我是该课程的推动者。2011年正值清华百年校庆，我是时任清华大学建筑学院院长，当时一项重要的工作是，抓住学校跻身世界一流契机，推进学院国际化平台建设。刘健老师当时担任院长助理、负责外事工作，于是，作为国际化平台建设重要内容的英文课程自然落在她的肩上。其次，我是该课程的牵线人。学校每年都会收到来自世界各地的捐赠。不过，指明要捐赠课程建设的并不多见。在此，要感谢总部位于日本东京的 One Asia Foundation 基金会，特别是基金会董事长佐藤阳二（Yoji Sato）先生。正是他的意愿和远见、基金会的慷慨赞助，从一个方面促进了这门英文授课的国际化专业课的设置和发展。当然，还要感谢

This is a professional course in terms of not only its subject but also its teaching in English, for which most of the students signing up came from the English Program of Master in Architecture at the School of Architecture, Tsinghua University. Although it was a small class composed of dozens of students, it represented a big world constituted by the knowledge of such disciplines as Architecture, Urban and Rural Planning, and Landscape Architecture. What is rare is that the instructors were mostly professors and experts from all over the world who have achieved great success in their respective academic fields.

This is the course of Asian Community Lecture Series: A Comparison of Architectural and Urban Culture between East and West which was presided over by Prof. Liu Jian, Associate Dean of the School of Architecture, Tsinghua University.

Prof. Liu always impresses me with her conscientiousness. She would often make long-term detailed schemes in advance. Through years of accumulation, the course contents are rich enough to be published as a professional book, and the publication of this book once again proves Prof. Liu's working style. I am here to give her a praise!

Actually, I have a little connection with this book. First of all, I am a promoter of this course. In 2011 when Tsinghua University celebrated its centennial Anniversary, I was the Dean of School of Architecture. An important task of mine at that time was to promote the School's internationalization by taking good use of the opportunity that Tsinghua strove to become a world-class university. Then, as Assistant Dean, Prof. Liu was responsible for the School's international affairs, thus the construction of English courses, which was an important part of the internationalization policy, naturally fell to her. Second, I was one of the go-betweens of this course. Tsinghua receives donations from all over the world every year, but there is hardly a donation dedicated to setting up a course. Taking that into consideration, sincere thanks should be firstly given to Mr. Yoji Sato, Chairman of One Asia Foundation headquartered in Tokyo, Japan, whose willingness and vision, together with the generous sponsorship of the Foundation, promoted the establishment

赞助基金的真正牵线人原清华大学副校长、时任清华大学教育基金会负责人的杨家庆教授和清华大学（香港特别行政区）教育基金会董事梁国栋先生，他们两位实际促成了该项课程赞助基金项目的运行。

小小课堂，大千世界，既体现了刘健老师课程的特色，也从一个侧面展示了清华大学建筑学院国际化教学所取得的成果。希望"中外建筑与城市文化系列丛书"的出版，能为中国建筑学、城乡规划学和风景园林学专业教育领域的国际化教学添砖加瓦。

and development of this internationalized professional course taught in English. Special acknowledgement should also be made for Prof. Yang Jiaqing, former Vice President of Tsinghua University and former Head of Tsinghua University Education Foundation, and Mr. Liang Guodong, Trustee of The Tsinghua University Education Foundation (HKSAR) Ltd., who promoted the running of the sponsorship fund for this course.

A small class miniatures a big world. It not only embodies the distinguishing feature of the course, but also, to a certain extent, shows the achievement of the School of Architecture, Tsinghua University in the field of educational internationalization. We hope that the publication of A Series on Architectural and Urban Culture of East and West will make some contributions to the educational internationalized of Architecture, Urban and Rural Planning, and Landscape Architecture of China.

目录 CONTENTS

1 在城市上建造城市：基于棕地开发的可持续城市空间发展之路 ... 1
BUILDING THE CITY ON THE CITY: A SUSTAINABLE WAY OF SPATIAL GROWTH THROUGH BROWNFIELD REDEVELOPMENT

2 1990 年代以来北京旧城更新与社会空间转型 ... 21
URBAN REGENERATION AND SOCIO-SPATIAL TRANSFORMATION IN BEIJING OLD CITY SINCE 1990S

3 风格主义新城镇 ... 43
MANNERIST NEW TOWNS

4 山地生态 ... 59
HILLSIDE ECOLOGIES

5 1950 年以来的中国城市演进之路 ... 73
URBAN FORMATION IN CHINA SINCE 1950 SEEN FROM AFAR

6 东亚城市的新兴建筑地域 ... 93
EMERGENT ARCHITECTURAL TERRITORIES IN EAST ASIAN CITIES

7 奇迹与地域 ... 115
MIRACLES AND REGIONS

8 立于分水岭还是汇入洪流？——全球化背景下风景园林东方空间观的反思 ... 133
AT OR IN THE CULTURESHED? A REFLECTION ON THE EASTERN SPATIAL PROPOSITION OF LANDSCAPE ARCHITECTURE IN THE CONTEXT OF GLOBALIZATION

9 一个建筑师关于文化与环境的可持续性的论述 ... 151
AN ARCHITECT'S DISCOURSE ON CULTURAL AND ENVIRONMENTAL SUSTAINABILITY

在城市上建造城市：
基于棕地开发的可持续城市空间发展之路

BUILDING THE CITY ON THE CITY: A SUSTAINABLE WAY OF SPATIAL GROWTH THROUGH BROWNFIELD REDEVELOPMENT

刘健
博士，副教授
清华大学建筑学院

Jian Liu
Associate Professor, Department of Urban Planning & Design
School of Architecture, Tsinghua University

中文翻译：刘健
Chinese translation by Jian Liu

1.1 我们的城市能否不断扩张发展？

自18世纪末开启工业革命以来，人类社会便开始了持续的城市化进程，并在两个世纪之后的新千年开启之际，迈入所谓的"城市社会"，即全球多数人口居住在城市地区，而非乡村地区。根据联合国的统计，2007年人类社会有史以来第一次迎来城市人口超出农村人口的事实，此后一直保持这个状态；2014年，全球54%的人口居住在城市地区；预计2050年，全球将有60%的人口居住在城市地区（UN, 2014）。回顾世界上高度城市化国家的城市化进程不难发现，在城市化加速发展时期，城市几乎无一例外地出现了持续的空间扩张；19世纪和20世纪的北美和欧洲城市如此，过去30年的中国城市也同样如此。

改革开放以来，中国在持续的社会经济发展推动下，进入城市化加速发展阶段；城市化率从1980年起始终保持1%的年均增长速度，并在2011年突破50%进入城市社会，2015年达到56.1%。考虑到中国庞大的人口基数，年均1%的城市化率增速意味着每年约有1400万人口从乡村进入城市，每年要以等同于新建14座百万人口大城市的城市建设规模，为这些新增城市人口提供住房、就业和服务。这足以说明中国城市在过去几十年里完成的天文数字量级房屋建设的合理性。1985年以来，中国城市平均每年新增竣工建筑面积11.5亿平方米，30年累计新增344亿平方米。既然无论人口增加还是建设增长最终都必然落在土地上，城市的空间扩张就成为不可避免。1996年全国666座城市的建成区面积总和为2.0万平方千米，2015年656座城市的建成区面积总和为5.2万平方千米，平均每年新增1500平方千米。如果将建制镇纳入计算，相关

1.1 Can our cities always expand?

Since the onset of the Industrial Revolution in the late 18th century, the world has seen a sustained process of urbanization throughout the 19th and 20th centuries and entered into the so-called Urban Society at the turn of the new Millennium, which is distinguished by the fact that more people reside in urban areas than in rural areas. According to the United Nations (UN, 2014), in 2007, it happened for the first time in history that the global urban population exceeded the global rural population and the world population has remained predominantly urban thereafter; there were 54% of the world's population residing in urban areas in 2014 and 66% of the world's population is projected to be urban by 2050. When looking back on the urbanization of the most urbanized countries in the world, it can be seen clearly a phenomena that, during the process of accelerating urbanization, cities are inevitable to undergo spatial expansion. This is the case for developed countries in North America and Europe in the 19th and 20th centuries, as well as for China in the past decades.

After its reforms and opening-up in the late 1970s, China has stepped onto the accelerating track of urbanization along with its sustained socio-economic development. Since 1980, it has seen an annual increase of urbanization rate of over 1%, with its urbanization rate exceeding 50% in 2011 and reaching 56.1% by 2015. Taking into account the total population of China, the annual urbanization rate increase of 1% implies a rural-urban migration of about 14 million people every year, which calls for large-scale urban constructions to provide with houses, job opportunities, and public services, i.e. an equivalence of building 14 mega-cities of one million people every year. This justifies the astronomical figures of floor area completion in Chinese cities in the past decades, with an annual increase of 1.15 billion m^2 since 1985 and a total accumulation of 34.4 billion m^2 in the past three decades. Now that both the demographic growth and construction increment have to be physically settled on land, spatial expansion appeared inevitable for Chinese cities, with the built-up area increasing yearly by 1,500 km^2 since 1996, from 20,214 km^2 for 666 cities in 1996 to 52,100 km^2 for 656 cities in 2015. The number will be doubled if the designated towns are

数字将成倍增长。对城市个体而言，平均建成区面积从1996年的30.4平方千米扩大到2015年的79.4平方千米，20年时间里几乎扩大一倍。再以北京作为特大城市的代表，其中心城的建成区面积在过去30年里呈现爆炸式增长，从1980年的不足240平方千米增加到2010年的约1400平方千米，平均每年新增37平方千米（图1）。上述数据无不展示出中国城市化发展的巨大成就，同时也暗示了无限城市扩张带来的一些问题和挑战。

从宏观角度看，中国在此过程中经历了耕地数量和质量的显著降低。从1996年到2005年的10年间，中国的耕地总量以平均每年8000平方千米的速度急剧减少，导致耕地存量一度接近18亿亩的"红线"；尽管2012年之后耕地总量重新回弹至20亿亩左右，但耕地质量的降低依然令人关注。在中国，特殊的自然和地理条件导致其绝大部分人口居住在东部数量有限的国土上；这种情况由来已久，在1935年由中国地理学家胡焕庸明确提出，故被称为"胡焕庸线"，即连接东北瑷珲（即今黑河）和西南腾冲的一条直线。根据统计，目前有超过96%的中国人口居住在胡焕庸线以东的土地上，面积仅占整个国土面积的36%。值得注意的是，这个人口高度集聚的区域同时也是中国高质量耕地资源密集的地区，城市化进程的加速发展及其带来的城市空间持续扩张导致大量耕地被占。尽管在统计上，这些被占用的耕地会以其他地区的土地开垦或复耕方式加以平衡，但耕地的产出能力显然不可同日而语。与此同时，中国的城乡差距也在不断加大，在越来越多的中国城市，尤其是特大城市，因为持续的人口增长和空间扩大而不堪重负时，也有越来越多的村庄因为持续的人口流失而趋于衰败甚至废弃。当然，生态环境质量的恶

also taken into consideration. The average built-up area for a single city also doubled in about twenty years, from 30.4 km^2 in 1996 to 79.4 km^2 in 2015. In the case of Beijing which can be taken as an example of mega-cities, the built-up area of its Central City expanded in an explosive way in the past 30 years, with an annual increase of more than 37 km^2, from less than 240 km^2 in 1980 to about 1,400 km^2 in 2010 (Fig. 1). All the data demonstrate the great achievements of China's urbanization on the one hand, while imply some challenging problems caused by unlimited urban expansion on the other hand.

From the macro perspective of the country as a whole, China has seen a remarkable decrease of farmlands in terms of both quantity and quality. During the ten years from 1996 to 2005, China's farmland area decreased sharply by nearly 8,000 km^2 every year, making the farmland reserve approach to the redline of 1.2 million km^2. Although the figure rebounded to more than 1.3 million km^2 after 2012, the quality degradation of farmlands remains a big concern of the society. Due to China's specific natural and geographical situations, it has been for a long time that most of its population reside on a small part of its territory in the east. This phenomena was firstly described by Chinese geographer, HU Huanyong, in 1935, with a line in his name which links Aihui (called as Heihe today) in the northeast to Tengchong in the southwest. According to statistics, today there are more than 96% of China's total population living in the area to the east of HU Huanyong Line, which accounts for only 36% of its total land area. As this area is also where most of China's best farmlands of high productivity are located, many farmlands were occupied and built-up due to the sustained urban expansion along with the accelerating urbanization. Although statistically they have been complemented by the farmland reclamation in other parts of China, the productivity is obviously far below that of the original ones. At the same time, China has seen an increasing enlargement of disparities between the urban and rural areas, with more and more villages becoming depopulated and dilapidated, or even abandoned, while more and more cites, especially the big ones, becoming over-loaded and over-concentrated. There have been also the cases of ecological and environmental degradation, in terms of air, water and soil, due to the unlimited and disordered urban expansion.

图1　1980—2000年北京中心城的空间扩张
（资料来源：清华大学人居中心，2001）

Fig.1　Urban expansion of Beijing from the 1980s to the 2000s
(Source: Center for Human Settlements, Tsinghua University, 2001.)

化，空气、水体和土壤的污染，同样不容忽视，尤其在人口大量聚集的东部地区。

从城市个体的微观角度看，当城市随着持续的扩张发展而变得越来越大时，也常常不得不面对一系列城市病带来的痛苦。就北京而言，这包括但并不限于：城市近郊的耕地彻底消失，绿色地显著减少，交通通勤的距离和时间不断增加，居住空间出现社会分化，难以获得经济适用的住房和公共服务，以及地方文化特色的消逝；凡此，无不对城市的宜居性和市民的幸福感造成严重影响。

1940年代中期，面对人类社会在进入20世纪以后出现的一系列急剧变化，特别是两次世界大战的严重破坏和经济大萧条的深远影响，西班牙建筑师、时任哈佛大学设计学院院长瑟特教授曾经通过一本专著大声疾呼"我们的城市能否生存下去？"（SERT，1942）今天，在世界倡导可持续发展、中国推进社会经济转型发展的时代背景下，前文述及的各种问题值得我们深入思考"我们的城市能否不断扩张发展？"不可否认，城市发展必然受到资源和环境条件的限制，因此针对上述问题的答案显然会趋于否定。鉴于中国人口巨大、资源有限、环境脆弱的特定条件，否定的答案尤其重要。然而，根据诺瑟姆的城市化理论（NORTHAM，1975），中国仍处于城市化进程的加速发展阶段，在正常情况下，没有任何力量能够影响其前进的步伐，直至进入稳定发展阶段。根据联合国开发署于2013年发表的研究成果（UNDP，2013），中国的城市化率将在2030年达到70%的水平，开始进入稳定发展阶段。这表明，在未来十多年里，中国仍将保持年均1%的城市化增长速度，每年仍将有约1400万人口从乡村进入城市。为了继续容纳数量庞大的新居民及其所需的新建

From the micro perspective of a single city, especially those becoming bigger and bigger along with sustained urban expansion, they have had to face a number of headache challenges. Taking Beijing as example, the challenges include but not limit to the complete disappearance of farmlands and remarkable reduction of green space at the urban fringe, the intolerant increase of commuting distance and commuting time, the emerging segregation of housing in spatial layout, the difficult access to affordable housing and public services, and the confusing identity of local culture, all of which have in consequence damaged both the livability of the city and the happiness of its residents.

Back in the mid-1940s, facing the radical changes of the world in the early 20th century, in particular the serious destruction of the two World Wars and the profound influences of the Great Depression, Spanish architect, José Luis Sert, asked the question of "Can Our Cities Survive?" in one of his manuscripts (SERT, 1942). Today, under the circumstances that the world is advocating sustainable development and China is standing at a critical transition of socio-economic development, all the problems mentioned-above warrant our deep thinking on the question of "Can our cities always expand?" Taking into consideration the undeniable restrains of resources and environment, the answer obviously tends to be negative. This is particularly significant for China in view of its specific conditions of huge population, limited resources and fragile environment. However, according to the theory of Ray M. Northam on urbanization (NORTHAM, 1975), China is still in the acceleration process of urbanization and nothing would stop its step until it enters the stage of stability. It was estimated by UNDP in 2013 that China's urbanization rate would reach 70% by 2030 (UNDP, 2013), which means, in the coming years, the annual increase of urbanization rate by 1% will continue, bringing a rural migration of about 14 million to cities every year. Thus, it will be for sure that Chinese cities will continue to see spatial growth to accommodate a large number of new residents and a large scale of new constructions, but not necessarily following the traditional spatial development mode of incremental. Instead, a new spatial development mode must be set up in correspondence to the restrains of resources and environment.

设，中国城市也将继续保持空间增长态势。但是，空间增长并不意味着一定沿袭传统的增量空间发展模式；恰恰相反，在新的形势下，需要考虑建立一种新的空间发展模式，以应对资源和环境的限制。

2012年中国提出"新型城镇化"战略，促进城市化转型发展，强调从经济导向转为人本导向，从数量导向转为质量导向，从环境漠视导向转为环境友好导向；在空间发展方面，则强调通过实施存量规划，促进城市空间发展从增量导向转为存量导向，即以对建设用地存量的高效利用取代建设用地供应的持续增长。这意味着，未来的城市发展不必像过去一样在一片处女地上拔地而起，而会越来越多地出现在城市中因为各种原因而未被充分利用的已有建设用地上，亦即所谓的"在城市上建设城市"的新型空间发展模式。总体来看，这种空间发展模式将主要涉及城市中的三类地区：一是可以重新开发利用的废弃棕地，二是可以更新改造的衰败街区，三是可以提高密度的低密度建成区。在三类地区中，棕地开发因在可实施性方面的优势而被予以优先考虑。

1.2 作为建设用地存量资源的棕地

在全球范围内，美国是第一个针对棕地给出官方定义的国家，1980年出台的《综合环境应对、补偿与责任法案》（又称CERCLA或巨额资金法案）将其称为棕地场所，并给出明确定义，此后，这一概念被加拿大、英国、法国、德国等其他工业化国家广泛接受。尽管各个国家对棕地的概念定义并不完全相同，但对棕地的基本特征则有比较共同的认识，即：棕地是指曾经被作为工业或者仓储

This legitimates China's policy of New Urbanization issued in 2012, which highlights the transition of urban development from economy-oriented to human-oriented, from quantity-priority to quality-priority, and from environment-regardless to environment-friendly. In terms of spatial growth, this policy is interpreted into the concept of inventory-based planning, which advocates that the future spatial growth of Chinese cities should be more inventory-oriented rather than incremental-oriented, focusing more on the efficient land-use of the existing construction land reserve, rather than the continuous incremental of construction land supply. It means that new urban development in future may not necessarily take place on virgin lands, as it usually did in the past, but more and more on the existing construction lands of a city which are not efficiently used due to certain reasons, a new development mode which can be called as "Building The City on The City". Generally speaking, three kinds of areas are mostly concerned by building the city on the city, i.e. the disused brownfields which can be redeveloped, the dilapidated urban areas which can be renovated, and the low-density built-up areas which can be further densified. And comparatively, priority should be given to brownfield redevelopment because of its advantages in feasibility.

1.2 Brownfieds as Construction Land Reserve

Worldwide, the US is the first country to give an official definition on brownfields, called as brownfield site in the *Comprehensive Environmental Response, Compensation, and Liability Act* of 1980, which is also known as CERCLA or Superfund. Afterwards, this concept has been widely accepted by many other industrialized countries, such as Canada, the UK, France, Germany, and so on. Although each country has its own definition on brownfields, a generalization can be made to describe its primary natures. Brownfields refer to the land properties that were previously used for industrial purposes or certain commercial uses, such as warehouses and railway yards, and became derelict and disused later on due to some reasons; they may be contaminated by hazardous wastes or pollutants and are often partly or completely built-up but not urbanized.

Historically speaking, brownfields are the result of the

和铁路站场等商业场所使用的土地资产，因为某些原因处于废弃状态，可能因为危险废弃物或污染物而处于被污染的状态，部分或者全部属于建成区但未被城市化。

以历史的眼光看，棕地的出现是世界经济从工业化向后工业化转型发展的必然结果，受到基于知识的新型产业的崛起、基于资源的传统产业的衰败以及全球化进程的综合影响。它最早于1960年代出现在工业化国家，直接起因是随着传统制造业生产被转移到成本更为低廉的其他国家和地区而出现的大量工厂停产倒闭，1970年代因其对环境和人体健康的负面影响而引起社会关注，进入1980年代又因其在城市土地利用方面的巨大潜力而成为社会热点。首先，棕地作为特定类型的城市建设用地，可重新用于任何适宜的城市建设，且其在许多城市的土地利用中仍然占有相当份额；其次，棕地在建设之初多处于城市边缘，后因城市空间扩张而被纳入城市地区，因此往往拥有良好的区位条件，不仅临近城市中心，而且多被建成的城市地区环绕（图2）；再次，棕地上多建有数量庞大的工业建筑物和构筑物，无论从美学还是实用角度看，都是拥有较高价值的丰富遗存；最后，因为棕地少有永久居民且产权关系相对单一，因此棕地开发可免于居民迁移和产权置换的烦扰。

上述优势使得通过棕地开发以获得更高土地利用效率成为可能。因此1980年代以来，最早面对棕地问题的工业化国家率先开展棕地开发实践，在不同空间尺度上采取不同方式进行探索。下文将重点分析其中的两个成功案例，一是德国鲁尔作为区域层面的实践案例，二是法国巴黎作为城市层面的实践案例。

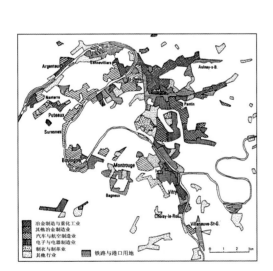

图2　20世纪初巴黎工业用地分布状况
（资料来源：Jean Bastié. Géographe Du Grand Paris. Paris: MASSON Editeur, 1984.）

Fig.2　Industries located at the urban fringe of Paris in the early 20th century
(Source: Jean Bastié. Géographe Du Grand Paris. Paris: MASSON Éditeur, 1984.)

world's socio-economic transition from industrialization to post-industrialization which is characterized by the rising of new knowledge-based industries and the declining of traditional resource-based industries. It firstly emerged in the industrialized countries in the 1960s, along with the shift of the traditional manufacturing to other regions or countries, under the circumstance of globalization, and the close down of the factories. In the 1970s, brownfields became a big concern of the society, because of its negative impacts on the environment and people's health through pollutions. Since the 1980s, brownfield redevelopment started to catch the attention of the society, because of its great potentials for urban land use. First of all, brownfield is a kind of constructible land, which can be reused for any appropriate urban constructions, and, in many cities, it still accounts for a considerable area of land use. Secondly, many brownfields have a favorable location in the city, usually not far away from the city center and surrounded by well urbanized areas, with thanks to their original location at the urban edge in the old days and the urban expansion afterwards which included them in a bigger urban area (Fig. 2). Thirdly, many brownfileds are equipped with a large number of industrial buildings and structures, which is a rich legacy of high value in both aesthetic and practical terms. Last but not the least, as brownfields normally accommodate no permanent resident and concern a unique property owner, brownfield redevelopment has less burden of relocating residents and dealing with property issues.

All these advantageous potentials make it possible to achieve a more efficient utilization of brownfields through redevelopment. Thus since the 1980s, the practice of brownfield redevelopment took place in some industrialized countries after the phenomena of brownfields firstly appeared in those countries. Various projects were carried out in different ways and at different levels, among which two successful cases are cited here, that is the brownfield redevelopment of Ruhr Region as a case at the region level and the urban regeneration of Paris Southeast as a case at the city level.

1.3 德国鲁尔：区域层面的棕地开发

德国鲁尔地区又称为鲁尔都市区，位于德国西部的北莱茵—威斯特伐利亚州，由分属3个行政管辖区的11个城市和4个县组成，并由位于西部的莱茵河、南部的鲁尔河、中部的埃姆歇河以及北部的里泊河相互联系在一起，土地面积4434平方千米，容纳人口约550万，是德国最大的城市聚集区，也是欧洲最大的都市区之一。

历史上，鲁尔地区在19世纪中期以后的一个多世纪里，一直扮演着德国工业发动机的角色，主要依靠煤炭开采和钢铁制造。该地区的采煤活动可追溯至14世纪，早期集聚在南部的鲁尔河谷，之后逐渐转移到中部的埃姆歇河流域。19世纪初开始工业化进程后，采煤业以及随之而来的钢铁制造业的大规模发展极大促进了鲁尔地区的人口增长和聚落发展，尤其在埃姆歇河流域，多特蒙德、埃森、杜伊斯堡、波鸿等重要城市发展迅速。此后至1970年代初，鲁尔地区一直保持着德国工业中心的地位。然而早在1950年代末，市场对煤炭和钢铁的需求量即大幅下降，鲁尔地区也由此开始进入产业结构调整。虽然采取了一系列应对措施，鲁尔地区仍在1970年代以后遭受了产业结构调整的重创，主要表现为大量煤矿和工厂关闭，相关生产被迁移至其他国家和地区，当地失业率居高不下。例如，1960年代末，鲁尔地区有26万工人从事采煤及其相关产业，而到了2006年，这一数据急剧下降到2.5万人。1975年鲁尔地区的失业人口为110万，2006年的失业率高达14%。

在大约20年时间里，就像其他发达国家的传统工业区一样，鲁尔地区沦为了一片巨大棕地：工业用地被弃

1.3 The Ruhr Region: Brownfield Redevelopment at Region Level

The Ruhr Region, Ruhrgebiet in German and known also as Ruhr Metropolis, is located in the federal State of North Rhine-Westphalia in the west of Germany. It is composed of 11 cities and four counties, which belong to three administrative districts respectively and are connected with each other by four rivers, i.e. the Rhine River in the west, the Ruhr River in the south, the Emscher River in the middle, and the Lippe River in the north. Covering an area of 4,434 km^2 and accommodating a population of about 5.5 million, it is the largest urban agglomeration of Germany, and one of the largest metropolitan areas of Europe as well.

Historically, the Ruhr Region had served Germany as industrial powerhouse for more than one century since the 1850s, mainly relying on coal mining and steel production. The coal mining activities started as early as in the 14th century, firstly in the Ruhr Valley and later northward to the Emscher Valley. After the region stepped into the period of Industrial Revolution in the early 19th century, the scaled development of coal mining, together with the following large-scale steel production, greatly promoted the demographic growth and human settlement development in this region, especially in the Emscher Valley where most of its cities concentrated and expanded to form an urban agglomeration, such as Dortmund, Essen, Duisburg, and Bochum, etc. Since then on, the region had kept its role of Germany's industrial center until the early 1970s, even though the market demand for coal and steel declined in the late 1950s when the region started to go through a profound industrial restructuring. Although a series of efforts was made to deal with the changes, the region was suffered from a serious consequence since the 1970s, which was characterized by the close of collieries, the shut-down of factories, and the shift of productions to other regions and countries, as well as a high level of unemployment. For example, there were about 260,000 workers working in coal mining and related sectors by the end of the 1960s, while the figure sharply dropped to only 25,000 around the year 2006; the unemployment was about 1.1 million in 1975 and the unemployment rate reached to 14% in 2006.

用，生态环境被污染，区域景观被破坏，烟囱、高炉和煤炭采掘机林立，但却毫无生息，处于非常艰难的状态（图3）。

幸运的是，鲁尔地区拥有一批无比忠诚的传统产业工人，他们因为鲁尔地区在一个多世纪里为他们及其家人提供了高质量生活而对此地怀有无比深厚的感情；面对困境，他们没有选择迁移至德国南部的新兴工业区，而是决定留在鲁尔地区，共同努力应对挑战。鲁尔地区的显著变化始于1989年实施的埃姆歇公园国际建筑展览项目。与传统的经济战略不同，这个为期10年的项目由北莱茵—威斯特伐利亚州政府发起，旨在通过建设一系列区域公园改变当地的环境面貌，并且通过建设一系列遗产之路促进新型产业，特别是文化创意产业的发展和聚集，包括工业文化之路、工业自然之路、工业建筑之路和地标艺术之路（图4）。

实质上，埃姆歇公园项目是一个典型的综合开发项目，涉及了棕地开发的方方面面。首先是有关水体和土壤的生态修复和环境治理，植物修复成为主要的应对手段；其次是经济发展，强调培养文化、创意、设计、展览、研究、咨询、服务和旅游等产业。为了配合达到上述两个核心目标，鲁尔地区在空间、社会和文化方面也采取了相应对策。在空间方面，曾经的工业用地被转而用于景观公园、科研开发、展览、休闲等用途，曾经的工业建筑物和构筑物经过改造之后成为博物馆、办公楼、展览厅、音乐厅、游乐设施（图5）；在社会方面，曾经的传统产业工人在经过必要的培训之后，重新就业从事服务业和文化旅游业，以自身的特有情感向来自国内外的游客介绍鲁尔地区；在文化方面，无论是有形的还是无形的工业遗产均得

For about two decades, the region was a huge brownfield site, just like the traditional industrial areas in other developed countries, with the industrial land abandoned, the ecological environment polluted, and the regional landscape degraded. Chimneys, furnaces and colliery machines were everywhere, yet mostly disused (Fig. 3).

All these consequences made the Ruhr Region in a very difficult situation at that time. However, the region is lucky to have such a faithful working class that is deeply attached to the region which provides them and their families with a high quality of life. Instead of migrating to the emerging new industrial districts in southern Germany, they chose to stay and make any efforts to change. The significant transformation of the region was remarked by the IBA Emscher Park Project launched in 1989. Quite different from traditional economic strategies, this ten-year project initiated by the Rhine-Westphalia state government aims at changing the physical image of the region through the construction of a series of regional parks, while promoting the growth and agglomeration of new industries, in particular the cultural and creative ones, through the establishment of a series of heritage trials, including the Industrial Culture Trial, the Industrial Nature Trial, the Industrial Architecture Trial, and the Landmark Art Trial (Fig. 4).

The IBA Emscher Park Project was actually a comprehensive project involving various aspects of brownfield redevelopment. Its primary concern was ecological restoration and environmental de-pollution of water and soil, for which phytoremediation was taken as a key measure. Its secondary concern was economic development which highlighted fostering the new industries of culture, creation, design, exhibition, research, consultation, service and tourism as well. Regarding the two key concerns, specific efforts were made in physical, social and cultural terms. Physically, the original industrial land use was transformed into other possible land uses, such as landscape park, research and development, exhibition, and recreation, with the original industrial buildings and structures being reused as museums, offices, exhibition centers, concert halls, and amusement centers, etc. (Fig. 5) Socially, many of the original industrial workers were reemployed after necessary training in the new industry of services, such as cultural tourism, to introduce their region with

图3 鲁尔地区一处尚未改造的棕地
（资料来源：刘健）

Fig.3 A brownfield site in the Ruhr Region
(Source: Jian Liu)

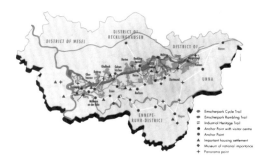

图4 埃姆歇公园国际建筑展项目
（资料来源：Kommunalverband Ruhrgebiet. The Ruhrgebiet-facts and figures. Essen 2004）

Fig.4 IBA Emscher Park Project
(Source: Kommunalverband Ruhrgebiet. The Ruhrgebiet-facts and figures. Essen 2004.)

到合理保护，成为发展文化旅游产业的重要资源，以及彰显地方特色的重要标志。

通过重塑区域景观、实施棕地开发，埃姆歇公园国际建筑展项目促成了鲁尔地区在空间形象和产业结构两个方面的成功转变，使当地许多废弃棕地得以重新高效利用，地区整体也从衰败的传统工业区华丽转身，成为充满生机的新兴文化产业基地，其中的关键在于在追求经济宏图的同时肩负起社会和环境责任。得益于埃姆歇公园项目的显著成效，埃森作为鲁尔地区的代表，获得"2010欧盟文化之都"的美誉。

1.4 巴黎东南区：城市层面的更新改造

巴黎——法兰西共和国首都，拥有2000多年历史的古老城市——是巴黎地区（又称法兰西岛大区Ile-de-France）的中心城市。巴黎地区作为欧洲最大的都市区之一，面积约1.2万平方千米，人口约120万。巴黎地区在19世纪初开始工业化进程，由此经历了持续的城市空间扩张，直至20世纪两次世界大战爆发。1859年拿破仑三世在位时，曾颁布法律明确巴黎的行政边界，确定辖区面积为105平方千米。至1929年，巴黎的人口规模达到历史最高的290万，行政辖区内除了森林、公园和绿地之外，几乎没有任何空地可以用于新的城市建设（图6）。有鉴于此，第二次世界大战结束后，巴黎将城市更新作为城市发展的主要模式。至1970年代，上述政策被进一步强化，以满足巴黎不断增长的城市建设需求，应对巴黎面临的诸多挑战，例如城市之间以及全球化进程中的国际竞争，在欧盟框架内争取欧洲首都的地位，以及社会经济转型带来

unique passion to the visitors coming both domestically and internationally. Culturally, the industrial heritage, both tangible and intangible ones, was reasonably preserved to become an important resource for cultural tourism development, as well as an unique icon of the local identity of the region.

By promoting the brownfield redevelopment through reshaping the regional landscape, the IBA Emscher Park Project successfully changed both the physical image and the industrial structure of the Ruhr Region, making many of the region's abandoned brownfields be reused more efficiently and turning the region from a typical traditional industrial area into a vibrant new cultural industry base. The key to its success is the due social and environmental responsibilities along with the economic ambitions. With thanks to the outstanding outcomes of this project, Essen, as a representative of the region, was nominated as European Capital of Culture 2010.

1.4 Paris Southeast: Urban Regeneration at City Level

The City of Paris, as the capital of France and a historic city of more than 2,000 years, is the urban core of Paris Region, known also as Ile-de-France, which is one of the biggest metropolitan areas of Europe, covering an area of over 12,000 km² and accommodating a population of about 12 million. Since the early 19th century when the region entered into the process of industrialization, Paris had seen a sustained urban expansion until the successive explosion of the two World Wars in the 20th century. By the year of 1929 when Paris saw the maximum population of 2.9 million in its history, there was almost no vacant land left for further urban constructions within its administrative boundaries, except the gardens, parks and forests, which were fixed in 1859 by Napoleon III to cover an area of 105 km² (Fig. 6). Thus, after the Second World War, Paris adopted the strategy of urban renovation as the key mode for its urban development. This policy has been further strengthened since the 1970s when Paris was facing an increasing demand for urban constructions to deal with a number of new challenges, such as the international competition among cities amid globalization, the potentiality of playing the role of the capital of Europe within the framework of

图5 北杜伊斯堡景观公园：从工业用地到休闲场所
（资料来源：刘健）

Fig.5 Duisburg-Nord Landscape Park: from an industrial land to a recreation area
(Source: Jian Liu)

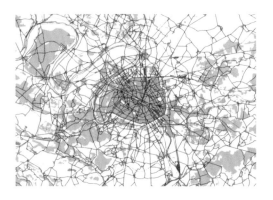

图6 1900年的巴黎城市空间发展状况
（资料来源：IGN et IAURIF. Trois siècle de cartographie en Ile-de-France (Volume 1). Les Cahiers de l'IAURIF. No. 119, Décembre 1997.）

Fig.6 Status quo of Paris' spatial development in 1900
(Source: IGN et IAURIF. Trois siècle de cartographie en Ile-de-France (Volume 1). Les Cahiers de l'IAURIF. No. 119, Décembre 1997.)

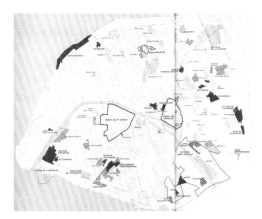

图7 1981年巴黎市内各类城市更新项目分部（红框部分为巴黎东南区）
（资料来源：APUR）

Fig.7 Different kinds of urban renovation projects in Paris by the year of 1981 (Paris Southeast marked by red line)
(Source: APUR)

的显著城市衰败。城市更新政策主要针对两类城市地区，即衰败的街区和废弃的棕地，其中多数分布在巴黎东部地区；这里曾经是传统产业最为集中的区域，同时也是受经济重构影响最大的区域，具体表现为工业用地和铁路战场被弃用，就业数量减少，人口结构老化，生活环境和城市景观破败。巴黎东南区则是其中的重要组成部分（图7）。

巴黎东南区位于巴黎东南、作为巴黎历史轴线的塞纳河两岸，距离巴黎历史中心不足3千米，土地面积为2.8平方千米，约占巴黎辖区面积的2.6%。早在17世纪，这里就是塞纳河水上货物运输的转运地，尤其以葡萄酒转运著名。19世纪中期，这里又建起两座重要铁路枢纽，即位于塞纳河右岸的里昂火车站和位于塞纳河左岸的奥斯特里兹火车站。因此，当地建有大量铁路站场和仓储货栈，但从1950年代开始，伴随铁路运输和水上运输的势弱而处于弃用状态。鉴于其在巴黎的重要区位及其在城市发展方面的巨大潜力，巴黎市政府在1970年代初即将巴黎东南区的更新改造纳入议事日程，并于1973年颁布《巴黎东南分区城市规划建设指导纲要》，从城市和区域两个空间层面出发，提出了巴黎东南区的发展目标，即：作为巴黎历史中心沿着塞纳河向东南方向的延伸，作为巴黎东部的新增长极与西部的德方斯地区相抗衡，一个容纳新兴产业的新型经济和文化中心，一个容纳居住、办公、商业、休闲、服务等多种城市功能的全新城市分区。1983年巴黎市颁布《巴黎东区发展计划》，列出未来6年将要展开的城市建设项目，其中包括位于巴黎东南区的若干协议开发区，例如塞纳河左岸的奥斯特里兹车站周边地区以及Tolbiac和Masséna等街区，塞纳河右岸的里昂火车站周

European Union, and the remarkable trend of urban declination due to the socio-economic transformation of the region, and so on. The policy of urban renovation mainly targeted at two kinds of urban areas, that is dilapidated neighborhoods and derelict brownfields, most of which were located in the eastern part of Paris where traditional industries were once concentrated and where the consequences of profound economic restructuring were mostly presented (Fig. 7), such as the abandonment of industrial lands and railway yards, the decrease of employments, the aging of demographic structure, and the degradation of living environment and urban landscape, etc.

Located in the southeast of Paris on both sides of the Seine River, the historical axis of Paris, Paris Southeast is only about 3 km away from the historic city center, covering an area of about 2.8 km^2 which accounts for 2.6% of its total area. Being the home of two important railway stations of Paris since the mid-19th century, i.e. Gare de Lyon and Gare d'Austerlitz, as well as the home for ports on the Seine River since the 17th century, it once accommodated a large number of railway yards and warehouses, both of which became dilapidated and disused since the 1950s along with the declination of railway and water transport. In view of its strategic location in Paris and its great potential for urban development, the urban renovation of Paris Southeast came into the agenda of Paris city government in the early 1970s. In 1973, the *Master Plan for Urban Development and Planning of Paris South-East Sector (Schéma Directeur d'Aménagement et d'Urbanisme du Secteur Sud-est)* was approved which oriented the urban renovation of Paris Southeast from both the city's and the region's perspectives. It was planned to be a southeastward extension of Paris' historic city center along the Seine River, a new growth pole in the east of Paris to balance La Défense in the west, a new economic and cultural centre characterized by new industries, and a new urban area accommodating multiple urban functions, such as residence, office, commerce, recreation, service, and so on. The key ideas of this plan were further developed by the *Programming Plan for Paris East (Plan Programme de l'Est de Paris)* issued in 1983, which listed the urban projects to be conducted in the coming six years. Since then on, the urban renovation of Paris Southeast has been carried out successively through the realization of a number

边地区以及Bercy和Chalon等街区，进一步落实了巴黎东南区城市更新改造的计划。

与鲁尔地区相比，巴黎东南区的一大优势是少有环境污染的困扰，因此其在废弃棕地上的城市更新主要集中在空间环境塑造以及对经济、社会和文化因素的考虑。空间方面，一是基于公交建设促进混合功能的土地开发，建设形成完整社区，这里共有5条地铁线、3条区域快轨线和4条郊区铁路线穿行，相互之间通过5个轨道交通站点实现换乘，极大地促进了高密度开发以及居住、办公、商业、娱乐、服务等多种城市功能的发展；二是通过公共空间建设，包括道路、广场、公园、广场、水岸等，塑造与周边地区相互协调的城市肌理。经济方面，强调通过开发建设产业园区，容纳诸如商务、商业、休闲、服务、科研、开发等新型产业，促进当地经济活力复苏。社会方面，一是强调通过多元化的住房建设，特别是社会住房与商品住房的混合开发，促进社会结构的稳定；二是强调通过建设包括图书馆、体育馆、高等学校、文化中心等在内的多种公共设施，提升当地的生活环境质量。文化方面，强调通过保护历史遗产及其要素以及重新利用古老的房屋和结构，继承和发展当地特色；例如，把曾经存放葡萄酒的仓储建筑改为餐饮、店铺，公园中开辟葡萄园以呼应与葡萄酒贸易相关的地方历史，在教育园地中设置小块绿地以反映当地作为13世纪巴黎郊区时的空间形态，等等。

自1980年代以来，遵循既定的发展战略，一系列城市建设项目在巴黎东南区得以实施，使其逐渐从废弃棕地转变成充满活力的城市新区，不仅实现了土地资源的高效利用，同时也取得了经济、社会、环境和文化方面的综合

of projects within the framework of Joint Development Zone (Zone d'Aménagement Concerté), including those of Austerlitz, Tolbiac, Masséna on the left bank (Paris Rive Gauche) and Bercy, Chalon, Gare de Lyon on the right bank.

Compared with Ruhr Region, Paris Southeast has an outstanding advantage that it was not suffered from any serious pollutions. Thus, its urban renovation on the derelict brownfields focuses mainly on reshaping the physical environment with due economic, social, and cultural concerns. It dealt firstly with the creation of complete communities through the mixed-use land development in coordination with the construction of public transport. There are five metro lines, three RER lines, and four regional railway lines passing through this area, being connected with each other through five interchange stations, all of which strongly support the high-density development of the multiple urban functions of residence, office, commerce, recreation, service, and so on. It dealt secondly with the configuration of urban tissue through the construction of public space, including roads, pathways, parks, squares, and water fronts. In terms of economy, it highlights regenerating the economic dynamic through the construction of new industrial areas which accommodate various new industries, such as business and commerce, recreation and service, and research and development, etc. In terms of society, it highlights stabilizing the social structure through the construction of diversified housing projects, in particular the integration of social housing with commercial housing, as well as upgrading the living environment through the construction of various public facilities, including libraries, stadiums, universities, colleges, culture centers, and so on. In terms of culture, it highlights developing the local identity through the protection of historic heritages and elements, as well as the reuse of old buildings and structures, such as the vine warehouses renovated into retails and shops, the vine yards memorizing the local history related to vine trading, and the educational garden in small parcels recalling Paris' suburban pattern in the 13th century.

Since the 1980s, with thanks to the orderly implementation of the plans in line with the prescribed strategies, Paris Southeast has been gradually transformed from a disused brownfield site into a dynamic new urban area, with a high land use efficiency

效益。尽管目前仍有部分项目在进行之中，但巴黎东南区已然在巴黎市和巴黎地区两个层面上发挥着经济和文化中心的重要作用，成为在城市上建设城市的成功典范（图8）。

1.5 北京的棕地开发实践

作为中国这个世界上最大发展中国家的首都，北京的棕地开发历史明显不同于鲁尔地区和巴黎。尽管拥有长达3000年的城市发展历史，但直到1940年代末，北京的城市建设主要集中在城墙环绕的62平方千米范围内。从1950年开始，在计划经济背景下，受到中国工业化进程的推动，北京城市空间开始缓慢扩张；1980年以后，在向社会主义市场经济转型发展背景下，受到中国城市化加速发展的强力推动，北京城市空间呈现了爆炸式的快速扩张。在1950年代，遵循"变消费城市为生产城市"的国家政策，北京在城市边缘地带规划建设了若干工业区，全部位于今天的三环和四环甚至五环路以外（图9）；直至1970年代末中国改革开放之时，除了一个规划预期之外的变化，北京的工业布局基本保持了上述特点。

1950年代末，为了响应国家的工业化号召，北京市曾颁布政策，允许在教育区和居住区发展非污染工业，结果导致工业企业在规划的工业区之外扩散发展。在随后10年间，由于缺乏基本的城市管理，工业无序扩张的状况进一步加剧，导致北京城市中心区出现大量小规模企业，污染企业与无污染企业兼而有之。工业企业的无序扩张对北京的城市发展产生了深远的负面影响。例如在土地利用上，相对居住用地和服务用地，产业用地比例不合理地偏

(a)

(b)

图8 巴黎东南区在棕地开发前（a）后（b）的对比
（资料来源：APUR, http://www.wikiwand.com/en/Seine）

Fig.8 Paris Southeast before (a) and after (b) urban renovation
(Source: APUR, http://www.wikiwand.com/en/Seine)

in both economic, social, environmental and cultural terms. Nowadays, although a number of projects are still going on, it has already played the role of economic and cultural center at both the city and region levels, proving the effectiveness of the development mode of building the city on the city (See Fig. 8).

1.5 Brownfield Redevelopment in Beijing

Being the capital city of China, the biggest developing country of the world, the story of brownfield redevelopment of Beijing is quite different from that of Ruhr Region and Paris. Although Beijing has a long history of over 3,000 years, its urban development remained mainly within its city walls by the end of the 1940s, which delimited an area of 62 km^2. It saw a slow urban expansion since 1950, driven by China's process of industrialization under the circumstance of planned economy, and then an explosive urban expansion since 1980, driven by China's accelerating process of urbanization under the circumstance of the transition to socialist market economy. In the 1950s, following the national policy of "turning consumptive cities into productive cities", a number of industrial areas were planned and built up in Beijing, all of which were located at the then urban fringe, beyond today's 3rd and even 4th or 5th ring-roads (Fig. 9). By the end of the 1970s when China initiated its opening-up and reforms, this was still the key pattern of Beijing's industrial layout, with only one unexpected change.

During the late 1950s, Beijing issued a policy to promote the construction of non-polluting factories in even educational and residential areas, leading to the disperse of industries beyond the planned industrial areas. In the coming decade further exacerbated the unreasonable disperse of industrial development due to the lack of regular urban management. A lot of small-scale factories, no matter polluting or not, were either inserted or built up throughout the central urban area of Beijing. This disordered disperse of industrial development brought about some far-reaching consequences to Beijing's urban development in the following years, such as the unreasonably high percentage of industry compared with that of residence and service in terms of land use, the functional conflicts between industrial production and daily life in terms of spatial layout, and

高；在空间布局上，工业生产功能和日常生活功能常常相互干扰；在环境问题上，工业污染不可避免地导致环境恶化。因此，1980年代初，当城市发展重新步入正轨，并面对产业结构调整问题时，北京开始对城市中心区的工业布局进行调整，重点就是那些土地利用效率低、空间布局不合理、环境污染严重的工业企业，以优化城市的用地结构和功能结构。这可以被视为北京棕地开发实践的开始，尽管相关工业用地并未处于废弃不用的状态。

在20世纪最后20年里，北京城市中心区约有150家工业企业被迁移或关闭，其中绝大多数是有污染的小型企业，释放出的工业用地被重新用于其他城市功能，例如新型产业以及相对匮乏的居住、商业、服务、绿地、公共设施等；与此同时，随着产业结构调整的不断深化，商务、金融、现代服务以及文化创意等新型产业迅速崛起（施卫良等，2010）。进入新世纪，出于经济和环境两方面的考虑，棕地开发成为北京城市发展战略的重要组成，并在2012年国家颁布新型城镇化战略之后得到进一步强化；无论在经济领域还是在空间层面，新兴产业逐渐取代传统产业的主导地位，促成了对位于城市边缘的大型传统工业区的大规模再开发。在21世纪的前10年，北京中心城有超过210家企业搬迁或关闭，释放出超过22平方千米的工业用地重新用于城市建设（施卫良等，2010），其中多数位于城市边缘的大型工业区，包括作为电子制造工业基地的酒仙桥地区，作为工程机械和纺织工业基地的通惠河两岸地区，作为化工基地的垡头地区，以及作为金属制造基地的石景山地区（图10）。因为各自产业结构不同，这些工业区的棕地开发也采取了不同方式，关注了不同问题。

the environmental deterioration in terms of industrial pollution. At the beginning of the 1980s, when Beijing resumed its urban development in a rational way and faced the new challenge of industrial restructuring, it had to deal with the problem of industrial dispersion in its central urban area, especially those factories of low land use efficiency, unreasonable spatial layout, and remarkable environmental pollution, in hope of optimizing the city's land use structure and functional layout. This can be regarded as the beginning of Beijing's brownfield redevelopment, even though the industrial lands at that moment were neither abandoned nor disused.

In the two decades up to 2000, about 150 factories in the central urban area of Beijing (SHI et al, 2010), mostly being small-scale and polluting, were either shut down or relocated, with the original industrial lands being transformed into other land use, such as new industry, residence, commerce, public facility, service, and green land. Meanwhile along with the proceeding of industrial restructuring, Beijing saw the rising of new industries in the late 1990s, such as business, finance, and modern service, as well as cultural and creative industries. Since 2000, brownfield redevelopment became an urban development policy of Beijing due to both economic and environmental considerations and its significance was further highlighted since 2012 under the circumstance of New Urbanization development. The new industries started to take the place of the traditional ones in not only economic but also spatial terms, leading to the redevelopment of the traditional industrial areas at the urban fringe. In the first decade of the new century, there were more than 210 factories shut down or relocated in the Central City of Beijing, releasing a land area of more than 22 km^2 for redevelopment (SHI et al., 2010), most of which are located in the large-scale industrial areas at the urban fringe, such as Jiuxianqiao as electronic manufacturing industry area, Tonghui Canal North-South as mechanical engineering and textile industry area, Fatou as chemical engineering industry area, and Shijingshan as metallurgy manufacturing industry area (Fig. 10). Because of the differences in industrial structure, these areas saw their brownfield redevelopment since 2000 concerning different aspects and done in different ways.

Talking about the brownfield redevelopment in Beijing, 798

图9　基于《北京市总体规划方案》1958年修订稿形成的北京工业空间布局（图中蓝色部分）
（资料来源：北京城市规划委员会，1958.）

Fig.9　Industrial layout in the urban area of Beijing proposed by the Preliminary Master Planning Scheme for Urban Construction of Beijing revised in 1958 (illustrated in dark blue)
(Source: Beijing Urban Planning Commission, 1958.)

关于北京的棕地开发，位于酒仙桥的798厂毫无争议是第一个成功案例。这家建于1952年的无线电设备国有企业，地处北京中心城东北郊，首都机场高速沿线、四环路和五环路之间，占地60公顷，拥有建筑面积23万平方米，是中国电子产业三个诞生地之一。在当时的社会主义联盟框架下，其规划建设得到来自前民主德国的大力支持，因此厂房建筑具有典型的包豪斯风格特点。经过1990年代末的产业调整，798变身为七星科技有限公司，原有车间被迁出，多数厂房被腾空。因其突出的建筑美学和实用价值，这些空置的厂房很快引起一群艺术家的关注。他们恰好因故刚从原住地迁出，正在寻找新的落脚之处，于是在2000年成为798最早的一批租户，在此设立了自己的艺术家工作室。之后，伴随文化创意产业在北京的迅速发展，以及中央美术学院在2001年搬迁至附近的望京地区，在产业集聚效应作用下，798很快吸引了涉及艺术、电影、媒体、设计、咨询、服务等诸多产业门类的400多家企业，至2008年已经成为享誉国内外的艺术区。显然，798的棕地开发主要是通过利用原有厂房容纳新型产业而实现土地用途的转变，并不涉及任何新的建设，也没有污染治理的困扰。

相比之下，位于北京中心城东部、长安街东延长线上、二环路和四环路之间的通惠河两岸地区的棕地开发起步更早、规模更大，且方式截然不同。这里主要汇集了建于1950年代的若干机械工程企业和纺织企业，连同周边的铁路站场和仓储货栈，占地达数平方千米。1990年代初，受产业结构调整影响，加之邻近规划中的北京CBD的特殊区位，这里的多数企业陆续搬迁或关闭，原有工业用地被转而用于发展商务、金融、居住、休闲、

Factory in Jiuxianqiao is always indisputably praised as the first successful case. Located in the northeast of Beijing Central City, along the Airport Expressway and between the 4th and 5th ring-roads, 798 occupies an area of more than 60 hm² and accommodates various kinds of buildings of 230,000 m² in floor area. Built up in 1952 as a state-owned company of wireless equipments, it was one of three birthplaces of China's electronic industry, physically characterized by the industrial buildings of Bauhaus style, which were designed by the architects coming from German Democratic Republic within the framework of socialist alliance. After the industrial restructuring in the late 1990s, it became Seven Star Science & Technology Co. LTD, with the production being relocated and most buildings being vacated. Very soon, its industrial buildings of high architectural quality in both aesthetic and practical terms attracted the attention of a group of artists, who were recently removed and looking for a new place to stay, and they became the first leaser of 798 in 2000 to set up their studios. Afterwards, with thanks to the rising of cultural and creative industry and the moving of China Academy of Fine Arts to Wangjing in 2001, 798 saw a remarkable effect of industrial aggregation, attracting more than 400 enterprises of arts, film, media, design, consultancy, and service, etc. and becoming a well-known art district in 2008 which enjoys a high reputation both at home and abroad. Obviously, the brownfield redevelopment of 798 mainly concerned the change of land use by reusing the existing buildings to accommodate new industries, yet without any new constructions and any necessity of de-pollution.

The brownfield redevelopment of Tonghui Canal North-South was carried out even earlier than that of 798, as well as in a larger scale and in a different way. In the east of Beijing Central City, on both sides of Tonghui Canal and along the eastward extension of Chang'an Avenue, between the 2rd and 4th ring-roads, a number of mechanical engineering and textile factories were built up in the 1950s, which, together with railway yards and warehouses, occupied a large area of several square kilometers. Since the early 1990s, under the circumstance of profound industrial restructuring and in view of the proximity to the planned Beijing CBD, many of the factories were either shut down or relocated, making it possible to redevelop the original industrial lands for

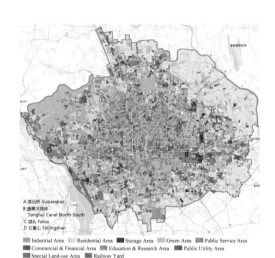

图10 21世纪初北京中心城工业用地布局（图中灰色部分）

（资料来源：北京城市规划设计研究院.《北京城市总体规划（2004—2020）》. 经作者改绘。）

Fig.10 Actual industrial layout in the urban area of Beijing in the early 2000s (illustrated in grey)

(Source: Beijing Municipal Institute of Urban Planning and Design. City Master Plan of Beijing 2004—2020. Amended by the author.)

服务等产业以及用于绿地建设，成功地将这个传统工业区的大部分地区转变成由若干办公区、居住区和休闲区组成的城市新区。目前，这里仍有数十公顷的铁路和仓储用地有待开发，考虑到周边汇集了中央电视台、北京电视台、中国传媒大学等重要媒体机构，规划将在此建设国家媒体广告园。和798相似，通惠河两岸的棕地开发主要涉及土地用途转变，以容纳新型产业，但却采用了截然不同的推倒重建的方法。截至目前，除原纺织厂的少量车间厂房被保留下来，并被改建成设计中心之外，原有的工业建筑物和构筑物全部被拆除，代之以全新的建筑。与798相似，通惠河两岸的棕地开发并未涉及污染治理问题，除了当地的污染问题并不严重外，当时整个社会对于环境问题关注不够也是其中原因之一。

相较于798和通惠河两岸，位于垡头的北京焦化厂和位于石景山的北京首钢的棕地开发则因环境问题而更富挑战性。北京焦化厂地处北京中心城东南、四环路和五环路之间，它建于1950年代，是当时中国最大的焦化企业，也是北京最主要的供气来源，占地约150公顷。出于环境原因，焦化生产于2012年停止，厂址将被规划建设成为由一座工业遗址公园、一个地铁站场和一个文化创意产业园区组成的多功能城市地区（图11）。按照规划，焦化厂的棕地开发将主要涉及土地用途转变和工业遗产保护，以容纳新型产业发展和新的城市功能。然而过去几年里，尚未有任何重要建设发生，相信未来在土壤污染问题得到妥善解决之前，也不会有任何大型建设展开。

首钢的情况也大致相似。首钢位于北京中心城西部、长安街的西延长线上，在五环路和永定河之间，占地约8平方千米，拥有职工约10万人。首钢创建于1919年，并

the new land use of business, finance, residence, recreation, service, and green space, etc., which successfully turned a large part of the traditional industrial area into a new urban area composed of business districts, residential communities, and recreation areas. Currently, there are still a land area of several decades of hectares left for redevelopment, occupied mostly by railway yards and warehouses, which is planned to be a National Media and Advertisement Park in view of the nearby congregation of the headquarters of China Central Television and Beijing Television and the Communication University of China. Similar with 798, the brownfield redevelopment of Tonghui Canal North-South mainly dealt with the change of land use to accommodate new industries, but was conducted differently in a bulldozer way. Except a small part of the textile workshops were preserved and transformed into a design center, the original industrial buildings and structures were totally demolished and replaced by new constructions. Also similar with 798, the brownfield redevelopment of Tonghui Canal North-South was not bothered by the issue of environmental pollution which was not serious in this case and was not socially concerned at that time.

Comparatively, the brownfield redevelopment of Beijing Coking Plant in Fatou and that of Beijing Capital Steel & Iron Group, or Shougang in short, in Shijingshan, both of which still remain in plans, are more challenging because of the environmental issue. Located in the southeast of Beijing Central City, between the 4th and 5th ring-roads, Beijing Coking Plant was built up in the 1950s as the biggest coking plant of China and the key gas supplier of Beijing, occupying an area of 150 hm^2. The production was stopped in 2012 due to environmental considerations and the site was planned to be redeveloped into a multi-functional urban area, composed of an industrial heritage park, a railway yard of subway, and a cultural and creative industry area. According to the plan, the brownfield redevelopment will focus on the change of land use and the preservation of industrial heritage to accommodate new industries and new urban functions (Fig. 11). However, nothing has happened and will happen before the serious soil pollution issue is effectively tackled.

The same for Shougang, too. Located in the west of Beijing Central City along the extension of Chang'an Avenue, between

图11 北京焦化厂棕地开发规划构想
（资料来源：北京华清安地建筑设计顾问有限责任公司）

Fig.11 A planning scheme for the brownfield redevelopment of Beijing Coking Plant
(Source: Beijng Huaqing Andi Architectural Design Institute.)

逐渐发展成为中国钢铁制造基地之一，其空间特点主要包括拥有大量工业遗存（如炼钢炉、储油罐、冷却塔、储煤仓等）、丰富建筑类型（如厂房、仓库、办公、宿舍、影院以及老的军事设备等）以及多种自然要素（如山体、河流、水面、绿化等）。出于环境保护和产业转型的考虑，首钢自2000年起开始外迁生产，至2010年彻底停产。早在首钢生产外迁之时，关于首钢棕地开发的规划研究即已展开，并已在北京市政府和企业之间达成一定共识，即首钢将被开发建设成为一个综合文化休闲区，承载金融商贸、科研开发、会议展览、商业、居住等不同功能，容纳工业主题公园、文化创意园区、行政管理中心、综合服务中心、文化休闲水岸、总部经济区域以及其他附属设施等七个功能分区。然而与焦化厂相似，因为存在污染，除了首钢二通因未受污染困扰而先期开发建设了中国动漫游戏城作为旗舰项目之外，首钢的棕地开发尚未有实质性建设。即便是作为旗舰项目的首钢二通中国动漫游戏城项目，尽管得到北京市政府针对文化创意产业的大力支持，在2011年启动之后也因为市场需求的不确定而进展缓慢。此外，考虑到首钢原本从事制造生产的庞大职工数量，工人的转岗就业也同样是一个巨大挑战。

由此可见，与鲁尔地区和巴黎相比，北京在产业调整过程中，并未经历棕地长期废弃之痛，并且几乎在同一时期开始了棕地开发实践。尽管北京早期的棕地开发实践并未过多关注社会、文化和环境问题，但当棕地开发成为城市发展的重要战略，且棕地开发项目规模越来越大，民众对社会公平、文化特色、环境健康问题越来越重视，社会、文化和环境问题就成为棕地开发不可回避的重要课题。

the 5th ring-road and Yongding River, Shougang was built up in 1919 and gradually developed into one of China's steel and iron production bases. Covering an area of about 8 km² and once holding an employment of about 100,000, it is physically characterized by rich industrial heritages (such as furnaces, oil tanks, cooling towers, and coal storage bunkers), colorful architectural types (such as workshops, warehouses, offices, dormitories, and old military facilities), and diversified natural elements (such as a hill, a river, large water surfaces, and quite good plantations). Due to environmental considerations and industrial restructuring, the production was removed out of Beijing since 2000 and completely stopped in 2010. The planning studies on the brownfield redevelopment of Shougang was initiated soon after the relocation of its production, achieving up-to-now a kind of agreement between Beijing Municipal Government and the enterprise that the site will be redeveloped into a comprehensive cultural and recreational district which will accommodate the activities of finance and business, research and development, conference and exhibition, commerce, and residence etc. in seven function zones, that is an industrial theme park, a cultural and creative industrial park, an administrative center, a service center, a cultural and recreational waterfront, a headquarter economy area, and other affiliation facilities. However and similarly, except the flag-ship project of China Animation & Game Town at the planned Shougang Ertong Industrial Park which is not suffered from any remarkable pollutions, no redevelopment project is carried out, mainly because of the environmental issue. Even the flag-ship project itself has seen a quite slow development since its initiation in 2011, in spite of the strong governmental support to cultural and creative industries, because of the uncertainty of the market demand. Moreover, taking into consideration the large number of employees originally engaged in manufacturing production, the reemployment is also a big challenge.

It can be seen that, compare with Ruhr Region and Paris, Beijing was not suffered from the long-time abandons or disuse of brownfields during the process of industrial restructuring and saw the initiation of brownfield redevelopment almost at the same time; although it was not concerned too much with the social, cultural and environmental issues in the early stage of brownfield

1.6 结论

城市化是人类社会发展的必然进程，在进入稳定发展阶段之前不会停止前进步伐；在城市化加速发展阶段，城市也必然经历持续的空间扩张。然而面对"城市能否不断扩张发展"这个问题，资源和环境制约无疑会使答案倾向于"不"；因此有必要对"在生地上建设城市"的传统空间发展模式进行反思，推广倡导建设用地循环利用的"在城市上建设城市"的新型空间发展模式。这一空间发展新模式主要涉及开发利用废弃棕地、更新改造衰败街区、提高低密度建成区的建设密度。其中，棕地开发因其独特优势和巨大潜力而被赋予优先考虑，例如棕地本身就是数量可观的建设用地储备，且在城市中区位优势明显，拥有高质量的丰富建筑遗存，可免于居民搬迁的困扰。

作为社会经济从工业化向工业化转型发展的必然结果，棕地不仅享有丰富的工业建筑遗存，同时也饱受一定程度的污染困扰。凡此使得棕地开发必然是一类综合开发项目，除了塑造空间形态之外，还将涉及生态修复、污染治理、社会重构、遗产保护、突出特色等问题。具体而言，在环境方面，重视利用植物修复等自然手段，去除水体和土壤的污染，重建当地的生态环境；在空间方面，重视将原本的工业用地转而用于其他任何可能的用途，将原有的工业建筑物和构筑物转而用于任何可能的功能，配套建设任何必需的公共设施，塑造全新的城市街区和城市肌理；在经济方面，重视通过产业调整，促进文化创意、高科技、现代服务等新型产业取代采矿、制造、重化工等传统产业；在社会方面，重视将传统的蓝领工人转变为手工业领域的新型灰领工人，甚至服务和管

redevelopment, it had to later on when brownfield redevelopment became an important urban development policy, the brownfield redevelopment projects became bigger and bigger, and the public was more and more conscious about social equity, cultural identity, and environmental healthy.

1.6 Conclusions

Urbanization is a natural development process of human society and no one can stop its pace until it reaches the stage of stability. Although cities are inevitable to undergo spatial expansion during the process of accelerating urbanization, the resource and environment restrains of a certain country or region will give the answer of "No" to the question "Can our cities always expand?" In that sense, the traditional spatial development mode of "building the city on virgin lands" should be retrospected and the new spatial development mode of "building the city on the city" which advocates the recycling use of constructible land reserve should be promoted. It may refer to the redevelopment of disused brownfields, the renovation of dilapidated urban areas, and the densification of existing low-density built-up areas, among which brownfield redevelopment is comparatively in priority because of its advantageous potentials, such as the considerable account in the existing construction land reserve, the strategic location in the city, the rich architectural legacy of high value, and the less burden of relocating residents.

Emerging as an inevitable result of socio-economic transition from industrialization to post-industrialization, brownfileds are specifically characterized by the contaminations and pollutions to some extent, as well as the legacy of industrial buildings and structures. All the facts make brownfield redevelopment a kind of comprehensive project that concerns not only the reshaping of physical environment, but also ecological restoration, environmental de-pollution, social restructuring, heritage protection, and cultural identification. Environmentally, it refers to the de-pollution of water and soil and the ecological restoration of the local environment, mostly in a natural way such as phytoremediation. Physically, it refers to the readjustment of the original industrial land use to any other possible land uses, the reuse of original industrial buildings and structures to any

理领域的现代白领工人，同时组建新型社区和新型社会关系；在文化方面，重视保护工业遗存、塑造地方特色。

面对"在城市上建设城市"的空间发展新模式，以及棕地开发的复杂难题，需要规划师建立长远的观点、动态的观点和综合的观点。长远观点要求把城市发展视为可持续的过程以及建设用地资源的循环利用；动态观点要求通过合理的基础设施和公共空间布局，建构一个灵活的空间框架，容纳城市发展的动态演进以及土地利用的可能变化；综合观点要求通过跨专业的合作应对棕地开发这样的困难课题。

other possible functions, the equipment of any necessary public utilities, and the creation of a new urban area and a new urban tissue. Economically, it refers to the restructuring of industries, with the traditional ones, such as mining, manufacturing, and heavy chemical engineering, being replaced by new industries, such as cultural and creative industries, high-tech industries, and service industries. Socially, it refers to the transformation of traditional blue-collars to new grey-collars for handcraft or modern while-collars for services and management, as well as the construction of new communities and the formation of new social structures. Culturally, it refers to the protection of the industrial heritages and the identification of the local culture.

Taking into consideration of the new spatial development mode of "building the city on the city" and the complexity of brownfield redevelopment, urban planners should at least build up the long-term view, dynamic view, and comprehensive view in correspondence. A long-term view means to view urban development as a sustainable process and constructible land as a kind of resources that should be used in a recycling way. A dynamic view means to structure a flexible spatial framework through the reasonable layout of infrastructures and public spaces which can function as a skeleton to accommodate the dynamic evolution of urban development and possible changes of land use. A comprehensive view means to rely on multi-disciplinary collaborations when dealing with the tough task of brownfield redevelopment.

参考文献 / References

LIU J, 2015. Transformation of Industrial Layout in the Spatial Planning of Beijing[M]//Michele Bonino and Filippo De Pieri. Beijing Danwei: Industrial Heritage in the Contemporary City, Berlin: Jovis Publisher: 56-73.

NORTHAM R M, 1975. Urban Geography. New Yokr, London, Sydney, Toronto: John Wiley & Sons, Inc.

SERT J L, 1942. Can Our Cities Survive? An ABC of urban problems, their analysis, their solutions. Cambridge, Harvard University Press; London, H. Milford, Oxford University Press.

United Nations Department of Economics and Social Affairs, Population Division, 2014. World Urbanization Prospect: The 2014 Revision, Highlights (ST/ESA/SER.A/352).

United Nations Development Program, 2013. China National Human Development Report 2013 - Sustainable and Livable Cities: Toward Ecological Civilization. Beijing: China Translation and Publishing Corporation.

和朝东，杨明，石晓冬，等，2014. 北京市产业布局发展现状与未来展望[J]. 北京规划建设，（1）：25-29.

克劳兹·昆斯曼，刘健，等，2007. 鲁尔传统工业区的蜕变之路[J]. 国际城市规划，22（3）：1-4.

李秀伟，路林，2011. 北京产业发展空间特征及利用策略[J]. 北京规划建设，（6）：53-56.

刘健，2007. 鲁尔区域特性之我见[J]. 国际城市规划，22（3）：60-65.

刘健，2007. 从衰败到复兴的蜕变——国外老工业区改造之经验启发[J]. 北京规划建设，（2）：21-27.

刘健，2013. 注重整体协调的城市更新改造：法国协议开发区制度在巴黎的实践[J]. 国际城市规划，28（6）：57-66.

施卫良，杜立群，王引，等，2010. 北京中心城（01-18片区）工业用地整体利用规划研究[M]. 北京，清华大学出版社.

1990 年代以来北京旧城更新与社会空间转型

URBAN REGENERATION AND SOCIO-SPATIAL TRANSFORMATION IN BEIJING OLD CITY SINCE 1990S

刘佳燕
博士，副教授
清华大学建筑学院

Jiayan Liu
Associate Professor, Department of Urban Planning
School of Architecture, Tsinghua University

中文翻译：刘佳燕
Chinese translation by Jiayan Liu

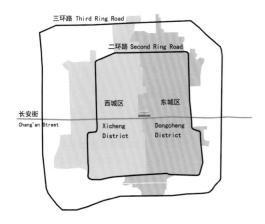

图1 北京旧城和东西城区的空间范围
（资料来源：刘佳燕）

Fig.1 Area scope of Beijing Old City, Dongcheng and Xicheng Districts
(Source: Jiayan Liu)

❶ 出于数据统计的原因，本文中部分数据对应的空间范围是"首都功能核心区"，包括东城区和西城区，占地92平方千米，2014年常住人口220万。

❶ For statistical convenience, some data used in this paper come from the statistical scope of the "Core Districts of Capital Function", which has a land area of 92 km² consisting Dongcheng District and Xicheng District, with 2.2 million permanent residents in 2014.

2.1 背景介绍

20世纪末期以来，在全球化、产业升级和区域竞争的背景下，城市更新在世界各地再次掀起热潮，并因其与全球资本和文化战略的结合，以及由此带来的绅士化、社会空间的极化与隔离等一系列社会问题而备受关注。在我国，自改革开放以来，特别是1990年代之后，各大城市纷纷致力于以旧城为核心的更新改造，但大部分工作都聚焦于物质性问题，包括风貌保护、住房改善、人口疏散、环境更新等，而社会性问题却往往被忽视，例如改造之后居民的分布和构成究竟发生了什么变化，是否出现了与西方类似的绅士化和社会空间隔离等现象，抑或还有哪些不同的特征，等等。随着中央提出"新型城镇化"的发展战略，越来越多的城市建设面临从外延式空间扩张向内涵式品质提升的转型，城市规划亦面临从空间蓝图向公共政策的转型，这些都要求我们跳出以往城市规划和建设中"见地不见人""见数量不见构成"的局限，真正贯彻落实"人本"思想。

北京旧城通常指城市二环路（原明清北京城城墙位置）以内的地域范围，共计62平方千米（图1）❶。北京旧城被我国著名建筑学家梁思成先生誉为世界"都市计划的无比杰作"，在20余年快速城市化和全球化浪潮下，经历了一次又一次更新改造的洗礼，也遇到了诸多的严峻挑战和问题。以下将对北京旧城1990年代以来的更新改造历程进行回顾，并围绕其社会空间转型的主要特征和问题进行探讨。

2.1 Introduction

Starting from the end of the last century, the world underwent a new wave of urban regeneration which attracted heated debates for its particular association with global capital and cultural strategies, as well as the consequent gentrification, socio-spatial polarization and segregation. In China, big cities have put considerable efforts into old city regeneration since the reforms and opening-up, especially after the 1990s. Most projects focused on physical aspects, such as housing and infrastructure improvement, population relocation, and historic townscape preservation, while social aspects were often overlooked, such as how urban regeneration influenced resident distribution and composition, whether China faced the problem of gentrification or socio-spatial segregation as Western countries did, and how China differed from the West in these aspects, etc. Guided by the "New Urbanization" strategy put forward by the Central Government in the 2010s, there have emerged in China two important transformations in the field of urban planning and construction, i.e. the shift of urban planning's attribute from physical planning to public policy and the transformation of urban construction from extensive expansion to quality-centered redevelopment. All requires a reflection on the previous development process through socio-spatial perspective to gain enlightenment to future urban regeneration strategies.

Beijing Old City usually refers to the area of 62 km² within the Second Ring Road, which was once closed by the Ming & Qing City Walls (Fig. 1)❶. It has a worldwide reputation of "an incomparable masterpiece of urban planning", acclaimed from LIANG Sicheng, one of the pioneers and founders of Chinese modern urban planning and architecture, and has a dramatic experience of urban regeneration over the past two decades in the context of rapid urbanization and globalization. A general review will be conducted on the urban regeneration of Beijing Old City since the 1990s, and the main features and problems in relation to its socio-space transformation will then be discussed.

2.2 Urban Regeneration of Beijing Old City since 1990s

After the founding of P. R. China in 1949, the building stock in Beijing Old City expanded from one million m² to over 30 million m² in the 1980s, by means of construction on vacant lands, plug-in construction and on-site reconstruction (DONG, 1998). However, housing supply did not catch up with the rapid growth of residents and building dilapidation. Moreover, poor infrastructure and illegal construction made housing issue even worse. In 1990, the net residential area per capita of urban residents in Beijing was only 11.17 m², slightly higher than that of 6.7 m² in 1978 (Beijing Municipal Bureau of Statistics and Beijing Survey Team of National Bureau of Statistics, 2009). Hereafter, along with the land and housing reforms, Beijing's urban fixed-asset investment and housing construction saw a rapid growth (Fig. 2) and Beijing Old City witnessed the thriving of urban regeneration to improve the residential conditions and update the industrial and social structures. Based on a historical review over the past two decades, the urban regeneration of Beijing Old City may be divided into three stages as follows.

2.2.1　1990—1997: Dilapidated housing redevelopment driven by real estate development

Prior to the 1990s in Beijing, within the framework of welfare-oriented public housing allocation system, most of the funding for urban housing construction came from governmental finance and there was always a serious shortage of funding facing the ever-increasing dilapidated houses and fast-growing urban residents. In the early 1990s, Beijing Old City had a large amount of dilapidated-housing floor area, i.e. 10 million m², occupying an area of over 1,900 hm² and housing a population of around 920,000 (Urban Construction Committee of Beijing CPPCC, et al., 2000). In the meantime, the city's rapid socio-economic development greatly propelled the urban regeneration of the Old City, because of a number of factors: first, rapid urbanization brought about a huge demand for housing; second, the unsound infrastructure, dilapidated buildings and overcrowded living environment in the Old City could hardly meet people's increasing living needs; third, as the Old City's economic structure shifted

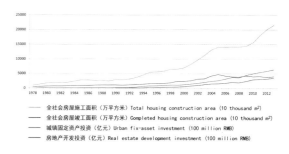

图2　1978—2013年北京城镇固定资产投资和房屋建设情况
（资料来源：数据来自北京市统计局，等，2014）

Fig.2 Fixed-asset investment and housing construction in the urban areas of Beijing during 1978—2013
(Source: based on Beijing Municipal Bureau of Statistics, et al., 2014)

需求；②旧城基础设施匮乏、建筑破败、过度拥挤的居住环境已经难以满足人们日益提高的生活需求；③旧城内以传统工业为主导的经济格局逐步向金融、商贸等第三产业升级，政府从招商引资的角度出发迫切需要改造旧城；④随着城市土地市场逐步形成，旧城土地价值凸显，成为投资开发的热点。

1990年4月，北京市政府第八次常务会做出加快危旧房改造的决定，确定了"一个转移"（城市建设的重点由新区开发为主转移到新区开发与危旧房改造并重）、"一个为主"（危旧房改造以区为主）和"四个结合"（危旧房改造与新区开发、住房制度改革、房地产经营以及保护古都风貌相结合）的改造方针。随后不久，北京市政府在东城区菊儿胡同、西城区小后仓等试点工程的基础上启动了大规模的危改工作，划出的第一批37个危改区中，22个分布在旧城。

随着《中华人民共和国城镇国有土地使用权出让和转让暂行条例》的颁布实施，房地产热潮席卷全国，北京旧城成为海内外房地产开发商的首选之地，加上北京市将危旧房改造的可行性研究方案审批权下放到各区县，危旧房改造在1994年步入顶峰。"1992—1994年底仅两年间，全市便有175片危改区立项改造，这个数字几乎是1991年的5倍"（方可，2000）。1995年原北京市建委主任王宗礼在全市危旧房工作会议上指出，1994年"是历年来危改工程建设速度最快、竣工小区最多的一年。全年完成危改投资30多亿元，相当于前四年的总和"（魏科，2005）。

由于短时间内需要的改造资金巨大，政府转向寻求市场资源。开发商作为重要的资金来源，进而在很大程度

its priority from traditional industries to tertiary industries (finance and business in particular), the government saw an urgent need to transform the Old City in order to attract more investments; finally, as the urban land market gradually took shape, the Old City became a hot spot, even at national level, for development investment for its highly potential land values.

In April 1990, the Standing Committee of Beijing Municipal Government proposed an important decision to facilitate the dilapidated housing redevelopment, in line with the following principles, that is redevelopment shifting its focus from new areas to both new areas and the Old City, district governments being the main body of redevelopment, as well as dilapidated housing redevelopment being integrated with new area development, housing system reform, real estate development, and historic townscape preservation. Soon after several pilot projects at Juer Hutong in Dongcheng District and Xiaohoucang in Xicheng District, Beijing Municipal Government launched the projects of large-scale dilapidated housing redevelopment. Among the first batch of 37 dilapidated house redevelopment projects, 22 were located in the Old City.

Meanwhile, there emerged a nationwide boom of real estate development soon after the promulgation of the *Provisional Regulations on Leasing and Transferring State-owned Land-use Rights of People's Republic of China* in 1990. Beijing Old City became a preferable choice for both domestic and overseas real estate developers and its dilapidated housing redevelopment reached a peak in 1994. "During the two years from 1992 to 1994, 175 dilapidated housing redevelopment projects were approved, almost five times of that in 1991" (FANF, 2000). In 1995, Wang Zongli, then Director of Beijing Construction Committee, pointed out in his report on Beijing Dilapidated Housing Work Meeting that, in 1994, over 3 billion RMB was invested into dilapidated housing redevelopment in Beijing, equaling to the total of the previous four years (WEI, 2005).

In urgent need for funding, the government turned to the market for help. It was in this circumstance that real estate developers became influential. Driven by enormous land value, the urban regeneration of Beijing Old City was gradually reoriented from dilapidated housing redevelopment to urban redevelopment that prioritized commercial facilities and office buildings. Within a few years, large-scale commercial projects sprung up in large

上影响了此阶段的改造进程。在旧城巨大的土地价值驱动下，改造形式逐步从危旧房改造演变成以酒店、办公等大型商业设施建设为主的城区再开发。短短几年时间，西单大街整体改造、王府井大街改造以及崇文门新世界等大量大型商业地产项目如雨后春笋般涌现。以坐落于东长安街1号的"东方广场"为典型，占地10公顷、总建筑面积约80万平方米，号称亚洲最大的商业建筑群之一❶。

据统计，1988—1998年底，全市"累计开工危改小区146片，竣工40片，竣工面积951万平方米，共动迁居民15.14万户，安置居民11.85万户，投入危改资金共计330亿元，拆除危旧房415万平方米"（朱自煊，2001）。大规模危旧房改造在改善旧城居民住房条件的同时，也暴露出一系列问题，具体如：①改造项目筹融资渠道单一，过度依赖开发商资金，难免导致经济利益至上的改造倾向，大量真正危旧的地区未得到有效改造；②改造项目普遍采取推倒重来的方式，造成拆迁规模过大、速度过快，因征地、拆迁引发的社会矛盾日益尖锐；③对历史地段的文化价值认识不足，致使许多传统风貌和特色景观遭到破坏；④过度扩张的房地产投资市场造成物价上涨及投资结构失衡；⑤大部分新建住宅为高端项目，相对而言大众急需的普通商品房供给不足。

针对以上情况，1995年和1996年国务院相继出台了《关于严格控制高档房地产开发项目的通知》（国发〔1995〕13号）和《国务院关于加强城市规划工作的通知》，加强了对高档房地产项目和城市规划工作的管控力度。受国家宏观调控影响，加上1997年下半年亚洲金融危机爆发，房地产呈现全国性的萎缩局面。北京也难以避免，大批靠引资启动的危改工程进入停滞期。

numbers, such as Xidan and Wangfujin Commercial Streets and Chongwenmen New World Centre. The Oriental Plaza, located on East Changan Avenue, occupying an area of 10 ha with a total floor area of around 800,000 m^2, was allegedly one of the biggest commercial building complexes in Asia❶.

Statistics show that, from 1988 to the end of 1998, Beijing had launched 146 dilapidated housing redevelopment projects, with 151,400 households being concerned and 118,500 being relocated. The projects involved an investment of 33 billion RMB and demolished some dilapidated houses of 4.15 million m^2 (ZHU, 2001). The large-scale redevelopment, while improving the residents' living conditions, revealed a series of problems notably. The limitation of financing channels resulted in an overreliance on real estate developers who excluded many real dilapidated areas from renovation for economic considerations. The widely adopted "demolition-reconstruction" mode resulted in large-scale and high-speed deconstruction, causing sharp social conflicts mainly during demolition and land acquisition; Due to underestimating the cultural values of historic sites, a large amount of historic buildings and sites were destroyed; The real estate investment market was overly expanded, leading to land price inflation and unbalanced investment structure; As most new houses targeted at high-grade market, there was a supply shortage of ordinary commercial houses.

Given these circumstances, China's State Council promulgated the *Notice on Strict Control of High-grade Real Estate Projects* (Guofa〔1995〕No. 13) and the State Council's Notification on Strengthening the Work of Urban Planning respectively in 1995 and 1996, which imposed stronger regulatory forces on high-grade real estate projects and urban planning work in general. When the Asian Financial Crisis broke out in the second half of 1997, China's real estate industry encountered a shrink and Beijing, of course, was not exempt. A large number of investment-driven redevelopment projects slipped into stagnation.

2.2.2 1998–2004: Large-scale Old City renewal driven by housing system reform and urban modernization

As the outcome of the previous stage, large-scale housing construction and commercial property development in Beijing Old City did not help to reduce population density, but on the

---

❶ http://www.orientalplaza.com/Menu/view/id/7.html

❶ http://www.orientalplaza.com/Menu/view/id/7.html

2.2.2　第二阶段（1998—2004年）：住房体制改革和城市现代化建设推动下的大规模旧城改造

前一阶段旧城内大规模的住宅建设和办公、商业地产的开发，不但没有降低人口密度，反而进一步增加了旧城的功能容量，导致原有道路和市政基础设施不堪重负。随着北京重塑古都风貌以及申办奥运等重大议题的启动，旧城改造的主题转向以道路和景观改造为代表的市政基础设施建设。例如1997年底的平安大街改造，以及之后的牛街拆迁和"两广"路拓宽等工程，主要依据的是1998年《北京市城市房屋拆迁管理办法》（市政府令第16号）中提出的以货币补偿为核心的危改政策。然而实际上，这种单纯依靠"市政带危改"的方式很快就因日益增大的资金和拆迁压力而举步维艰。

如果说旨在提升旧城环境品质的城市现代化建设为这一轮旧城改造的兴起创造了契机，那么真正为其注入源源动力的则是国家住房体制改革。针对1997年出现的经济疲软现象，国家出台了一系列积极的财政政策和调控措施，旨在刺激投资需求和鼓励消费，拉动经济增长，其中的一个核心领域就是住房体制改革。1998年7月国务院颁布《关于进一步深化城镇住房制度改革、加快住房建设的通知》，宣布从当年下半年开始，全国城镇停止住房实物分配，实行住房分配的货币化。2000年3月《北京市加快城市危旧房改造实施办法（试行）》（19号文）出台，提出了"政府组织扶持、居民自主决策、房改带动危改、解危适度解困"的危改新思路，并决定在"三区五片"（龙潭西里、金鱼池、天桥、右外西庄三条和牛街二期）进行试点，将过去福利制的危房改造变为政府、单位、个人三方共同出资的"危改加房改"，为长期制约旧城改造

contrary, attracted more functions and population, leading to the serious overload on existing roads and municipal infrastructure. As major projects were launched (e.g., to reshape the cityscape of the ancient capital and to bid for the Olympic Games), the urban regeneration of Beijing Old City shifted its focus to improving municipal infrastructures. Examples include the beautification of Ping'an Avenue, the redevelopment of Niujie area, and the broadening of Guang'an Avenue, all of which were in line with the monetary compensation policy for dilapidated housing redevelopment as proposed by the Administrative Measures of Urban Housing Demolition and Relocation in Beijing (The Municipal Government Decree No. 16) issued in 1998. However, this mode of dilapidated housing redevelopment driven by infrastructure improvement appeared hardly to go further under the increasing capital and demolition pressures.

While urban modernization aiming at upgrading physical environment quality propelled the transformation of Beijing Old City, the housing system reform constantly injected motive force into it. To cope with the nationwide economic weakness since 1997, China's Central Government introduced a series of effective fiscal policies and control measures, aiming to activate investment, encourage consumption, and stimulate economic growth. Among them, one core measure was housing system reform. In July 1998, the State Council released the *Notice on Further Deepening Urban Housing System Reform and Accelerating Housing Construction*, which proposed to replace the old in-kind housing allotment system with a new monetary and market-oriented one from the second half of the year. In March 2000, the *Measures of Accelerating the Implementation of Urban Dilapidated Housing Redevelopment in Beijing* (*Trial*) (Document No. 19) was promulgated, proposing a new thought of "organization and support by government, decision-making by residents, and dilapidated housing redevelopment and rehabilitation and moderately eliminating the case of housing difficulty through housing system reform." Five pilot projects in the Old City were decided, i.e. Longtanxili, Jinyuchi, Tianqiao, Youwai Xizhuang Santiao, and Phase II of Niujie. This was to turn the former welfare-oriented dilapidated housing redevelopment into "dilapidated housing redevelopment + housing system reform" invested jointly by the government, danwei (work units) and individuals,

的资金困境带来新的转机。北京市"十五"计划提出"全市危房改造工作以旧城区和关厢地区为重点,5年改造危房300万平方米。到2005年,基本完成城区现有危旧房改造。"危旧房改造成为北京市区政府重点推进的工作任务,"每年,各区都与市政府就将要实施的危改项目签订目标责任书,因此,各区领导便不遗余力地将加快危改作为唯一的目标,危改也成为其最为重要的政绩之一"(周乐,2002)。

与此同时,结合申办奥运的契机,北京市加大了古都风貌保护和整治力度。2000年和2003年,北京市相继投入3.3亿元和6亿元实施修缮保护计划和"人文奥运文物保护计划"(郑珺,2010),推进文化古都保护和旧城改造工作的结合。

借助住房体制改革释放的市场动力,通过市政带危改、开发带危改、文物保护带危改及19号文带危改等多种形式,北京旧城进入了前所未有的危旧房改造高潮,甚至大幅超越了上一阶段。2001年危旧房改造规模达到顶峰,全年拆除危旧房183.9万平方米(魏科,2005)。根据《东城区"十一五"时期城市建设发展规划》,2001—2005年间,(原)东城区"共完成危改投资119.2亿元,是前10年投资总额的5.2倍。累计动迁居民42000多户,是前10年动迁居民的2.3倍。五年共实现竣工面积200万平方米,是'九五'期间的5.3倍"❶。

这一阶段旧城改造的一大突破在于多途径解决改造资金问题。一方面,市、区政府的经济实力已有较大发展,并拓展了银行贷款、土地出让金、股票和基金等多样化的融资渠道;另一方面,广大居民经济实力不断增强,加上个人住房信贷的发展,促进了货币安置、就地回迁、异地

in order to eventually ease the long-time funding shortage for housing construction. As proposed by the 10th Five-Year Plan of Beijing, the dilapidated housing redevelopment in Beijing should focus on the Old City and its outskirts, redeveloping a total floor area of 3-million-m^2 in the following five years and eliminating all dilapidated houses in the urban areas by 2005. Dilapidated housing redevelopment thus became a priority in the urban district governments' agenda. "Every year, each district government would sign a goal-responsibility agreement with the municipal government on the dilapidated housing redevelopment projects to be implemented. Under this circumstance, district officials spared no effort to accelerate the redevelopment, which was also one of the most important indicators of their performance" (ZHOU, 2002).

In the meantime, along with its endeavor to bid for the 2008 Olympic Games, Beijing strengthened its urban renovation and historic preservation. In 2000 and 2003, Beijing Municipal Government invested in succession 330 million and 600 million RMB for the Old City renovation and preservation plan and the Cultural Heritage Preservation Plan of "People's Olympics" (ZHENG, 2010), so as to promote the integration of historic preservation and urban regeneration of Beijing Old City.

Under the market forces released by the housing system reform, Beijing Old City entered an unprecedented period of dilapidated housing redevelopment which was jointly driven by infrastructure improvement, real estate development, cultural heritage protection and Document No. 19. It exceeded the previous stage in terms of scale and came to a peak in 2001 when the annual demolished floor area reached 1,839,000 m^2 (WEI, 2005). According to *the Urban Construction and Development Plan of Dongcheng District during the 11th Five-Year Period*, from 2001 to 2005 in the former Dongcheng District, the dilapidated housing redevelopment attracted a total investment of 11,92 billion RMB, 5.2 times of the total over the previous decade, relocated over 42,000 households, 2.3 times of the total over the previous ten years, and completed a new floor area of 2 million m^2, 5.3 times of that in the 9th Five-Year period❶.

A breakthrough of this stage is the seek for redevelopment funds through multiple channels. On the one hand, with thanks to the sustained economic growth, both the municipal and district governments expanded and diversified the financing channels

---

❶
2010年北京市进行行政区划调整,原东城区和崇文区合并组建新的东城区,原西城区和宣武区合并组建成新的西城区。数据来自:北京市规划局. 东城区"十一五"时期城市建设发展规划[EB/OL]. 东城规划分局网站. [2015-03-31]. http://ldj.bjdch.gov.cn/n5687274/n5723305/n5738209/n5739298/8077185.html.

❶
In 2010 Beijing readjusted the administrative division, combining the former Dongcheng District with Chongwen District into new Dongcheng District, and the former Xicheng District with Xuanwu District into new Xicheng District. Data comes from: Beijing Planning Bureau. City Construction and Development Planning of Dongcheng District During the 11th Five-Year Period, Website of Dongcheng Planning Branch, accessed on March 31st, 2015 via: http://ldj.bjdch.gov.cn/n5687274/n5723305/n5738209/n5739298/8077185.html.

安置等多种拆迁安置办法的实行，使得来自单位和个体的社会资本成为改造资金的重要组成部分。以2002—2003年实施的南池子试点改造工程为例，涉及拆迁居民1076户，其中回迁居民300户，异地安置和货币补偿的776户（冯熙，2014）。在项目总支出的3.01亿元中，政府直接投入资金0.52亿元，其余的2.49亿元来自部分土地转让和向居民售房等（邰磊，2010）。

这一轮轰轰烈烈的旧城改造运动在大幅改善旧城风貌和居民住房条件的同时，也引发了诸多矛盾。随着危旧房改造的任务下放到区县一级，更新改造成为各区县政府的考核任务，也成为政绩工程，过度拆迁和突破控规等现象屡屡发生。1990—2003年间，北京共拆除胡同639条，是前40年的3.1倍(1949—1989年拆除胡同199条)（魏科，2005）。更严重的是，大规模拆迁导致不少特色历史文化街区迅速瓦解、消亡，甚至引发社会冲突。以牛街地区的危旧房改造为例。这里是全国闻名的回民聚居区，1996年危改启动，被誉为当时北京市拆迁面积最大、拆迁户最多、少数民族比例最高的危改区。大部分的回族居民不愿意远离清真寺，因而选择了回迁。改造后的牛街危改小区内，矗立起一座座现代主义风格的高层塔楼，紧紧包裹住中间保留下来的牛街清真寺，街区的传统历史风貌丧失殆尽。

2004年中央政府再次出手，推行"管严土地、看紧信贷"等宏观调控手段以应对全国性的房地产投资过快增长等问题。北京市政府结合中央有关北京历史名城保护的指示精神，反思并转变了旧城改造的工作方式。

through bank loan, land leasing fees, stock and others. On the other hand, the constant improvement of residents' economic conditions, together with the development of individual housing credit, facilitated the implementation of several resettlement measures, including monetary compensation, moving-back resettlement, and elsewhere relocation, etc., which highlighted the funds from individuals and danweis as an important source for redevelopment. Take the Nanchizi pilot project launched during 2002 to 2003 as example. It involved 1,076 households, among which 300 moved back for resettlement and 776 accepted either elsewhere relocation or monetary compensation (FENG, 2014). Among the total project expenditure of 301 million RMB, 249 million was raised through land leasing and house selling, whereas governments' direct investment was only 52 million (TAI, 2010).

While the ambitious urban renewal movement substantially improved the Old City's outlook and housing condition, it sparked a lot of controversies. As the mission of dilapidated housing redevelopment was downwards to the district level, it was more and more regarded as government officers' achievement to evaluate their performance. This, to a large extent, resulted in excessive forced demolitions and relocation, as well as frequent violations against regulatory plans. Statistics show that, during 1990 to 2003, there were in total 639 hutongs demolished in Beijing Old City, which was over 3 times the number of the previous 40 years when 199 hutongs were demolished during 1949 to 1989 (WEI, 2005). Moreover, large-scale demolitions led to rapid disintegration or disappearance of many featured historic and cultural areas. Sometimes they even caused social conflicts. The Niujie area for example is a well-known Muslim area for Chinese Hui people. In 1996 when the redevelopment project was launched, it was viewed as one of the most famous and sensitive dilapidated housing areas, with the largest floor area of demolition, the largest number of removed and relocated households, and the highest ratio of minorities ever in Beijing. Unwilling to reside far away from the mosque temple, most residents chose to move back. As a result, vast modern high-rise apartments erected on the site after the redevelopment, surrounding the only left and well-preserved mosque in the center, which deprived the area of its original historic feature.

Until 2004 when the Central Government put into effects the

2.2.3 新时期（2005年至今）：旧城有机疏解和整体保护思路下的新型更新模式探索

《北京城市总体规划（2004—2020年）》的出台推动了从区域协调的角度实现旧城保护与新城开发的统筹发展，强调通过"旧城有机疏散、市域战略转移"等措施，疏散旧城人口压力，保护其历史风貌特色。随后相继颁布的一系列规划和条例，包括《北京历史文化名城保护条例》《北京市"十一五"时期历史文化名城保护规划》《北京市文物建筑修缮保护利用中长期规划（2008—2015）》《北京旧城房屋修缮与保护技术导则》等，逐步构建起以旧城整体保护为核心，从文物保护单位、历史文化保护街区到历史文化名城的系统性的名城保护格局。

由此，强调旧城整体保护，推进小规模、渐进式的更新模式开始逐步成为社会共识。北京市政府提出"修缮、改善、疏散"的总体思路，确定"政府主导、财政投入、居民自愿、专家指导、社会监督"的方式，开展以"院落微循环改造""政府组织拔危楼""街巷胡同环境整治""六片文保区试点"为代表的"点、线、面"相结合的旧城保护和改造实践。2007年全市改造涉及旧城44条胡同、1474个院落、9635户居民，2009年因旧城区房屋保护修缮而疏散居民数降至2004户（北京市住房和城乡建设委员会，2008，2010）。

以大栅栏地区有机更新项目为例。作为北京旧城内保留相对完整的历史文化街区之一，曾经繁华的大栅栏地区面临人口密度过高、公共设施条件滞后、区域风貌不断恶化、产业结构升级困难等发展困境。2011年北京市政府和西城区政府启动大栅栏地区更新计划，成立北京大栅栏投资有限责任公司作为区域保护与复兴的实施主体，通过

macro control measures of tightening loans and strengthening land control to cope with the excessive growth of real estate investment, Beijing Municipal Government decided to change the way of Beijing Old City's urban regeneration in accordance to the Central Government's instruction on historic protection of the Old City.

2.2.3 2005 to now: Exploring a new urban regeneration mode under the thought of organic decentralization and overall protection

In order to promote coordinated regional development, *Beijing City Master Plan (2004–2020)* issued in 2004 highlighted the coordination of old city protection and new district development, aiming at easing the population pressure on the Old City while protecting its historic features by strategically decentralizing functions and population. A series of plans and regulations were promulgated in succession afterwards, including *Regulations on Historic and Cultural City Protection of Beijing, Historic and Cultural City Protection Plan of Beijing in the 11th Five-Year Period, Medium- and Long-term Plan for Historic Building Renovation, Preservation & Utilization of Beijing (2008–2015)*, and *Technical Guidelines for House Renovation and Preservation in Beijing Old City*, all of which built up a historic protection system centering on the Old City while covering cultural relics protection units, historic and cultural preservation blocks, and the historic and cultural city as well.

Since then the society has gradually reached a consensus on the small-scale and organic regeneration mode emphasizing the overall protection of Beijing Old City. Beijing Municipal Government put forward the general idea of "renovation, improvement, decentralization", developed the working route of "government's leadership, fiscal investment, residents' willingness, experts' guidance, community's monitoring", and identified multiple patterns of old city protection, including micro-rehabilitation of single courtyard house, government-led demolition of dilapidated houses, environment remediation of alleys and hutongs, and pilot projects in six historic and cultural blocks. According to Beijing Construction Statistical Yearbook, in 2007, the urban regeneration projects in the Old City involved 44 hutongs, 1,474 courtyard houses, and 9,635 households. Later in 2009, there were only 2,004 households decentralized from the Old City for house protection and renovation (Beijing Municipal Commission of

制定"区域系统考虑、微循环有机更新"的整体策略，推进"在地居民商家合作共建、社会资源共同参与"的改造模式。具体举措包括：①完善道路和市政基础设施建设，提升胡同和外部空间环境品质；②搭建开放的工作平台"大栅栏跨界中心"，作为政府与市场的对接平台；③结合每年的北京国际设计周举办"大栅栏新街景"设计之旅，吸引中外优秀的设计和艺术创意项目进驻老街区，在尊重街区原有肌理的前提下，通过设计的力量引入新业态，探索老房新用的多种途径；④设立"大栅栏领航员试点计划"，针对旧街区改造过程中的难题，面向社会广泛征集创造性思维和设计，推进长期性的合作共建实践❶。在杨梅竹斜街试点区范围内，通过货币和定向安置房对接补偿的方式，已有近600户居民实现了分批小规模自愿腾退，占需腾退总户数的1/3；街道公共环境和市政设施条件大幅改善；一批设计团队和创作室悄然入住，给胡同生活带来新的气息。

但另一方面，在日益激烈的全球化竞争下，各区县竞相打造品牌街区，一个个现代化园区建设计划在旧城内诞生，从东城区的"龙潭湖体育产业园"到西城区的"天桥演艺园区""金融街（拓展计划）"。西城区"十二五"规划中提出重点打造金融街、德胜科技园区、广安产业园、什刹海历史文化保护区、阜景历史文化街区、琉璃厂艺术品交易中心区、天桥演艺园区、西单商业区、大栅栏传统商业区和马连道茶叶特色商业区等十大功能街区，其面积总和占全区面积的44.1%❷。以其中的金融街（扩展计划）为例，在原金融街规划范围1.18平方千米的基础上，向西、南、东拓展至占地面积2.6平方千米，未来将进一步对接南部的广安产业园，形

❶ 参见"大栅栏"网站，网址：http://www.dashilar.org/。

❶ From the official website of Dashilar: http://www.Dashilarr.org/.

❷ 参见"北京西城"网站，网址：http://www.bjxch.gov.cn/。

❷ From the official website of Xicheng District of Beijing: http://www.bjxch.gov.cn/.

❸ 参见"北京金融街投资（集团）有限公司"网站，网址：http://www.fsig.com.cn/。

❸ From the website of Beijing Financial Street Investment Group Co. Ltd. http://www.fsig.com.cn/.

Housing and Urban-rural Construction , 2018, 2010).

Take Dashilar area as example. As one of the famous traditional commercial centers and the best preserved historic and cultural blocks in Beijing Old City, the once-prosperous Dashilar area has faced for decades the increasing development dilemmas, like high population density, backward public facilities, deteriorating landscape, and industrial restructuring difficulties. In 2011, Beijing Municipal Government and Xicheng District Government jointly launched the urban regeneration plan named "Dashilar Project" and established Beijing Dashilar Investment Limited as the principal actor for the area's preservation and revitalization, promoting the organic regeneration mode of "co-construction between local residents and entrepreneurs, joint participation of social resources". Specific measures include: upgrading the road system and municipal infrastructures to improve the quality of hutongs and exterior environment; building a shared network of "Dashilar Platform" to bridge the government and the market; holding the annual design activity of "Dashilar Alley" in collaboration with Beijing Design Week to introduce exceptional design and art projects into the old district and explore multiple routes to re-use old houses in a new way via design, while respecting the district' original texture; launching the "Dashilar Pilot" as a long-term investment in creative projects to explore the possibilities of a wide range of collaborations between architects, designers, artists, local residents and business owners❶.Within the pilot area of Yangmeizhu Xiejie, nearly 600 households have been relocated voluntarily through monetary compensation and resettlement, accounting for 1/3 of the total households defined by the relocation plan; the street environment and municipal facilities have been greatly improved; the settlement of a growing number of designers and studios have brought new vitality to the hutong life.

On the other hand, under the increasing competition brought by globalization, districts competed to build brand blocks, stimulating the construction of modern industrial parks in Beijing Old City. Examples include Longtanhu Sports Industrial Park in Dongcheng District, as well as Tianqiao Performance Area and Financial Street (expansion plan) in Xicheng District. The 12th Five-Year Plan of Xicheng District proposed developing 10 major functional areas including Financial Street, Desheng Science Park, Guang'an Industrial Park, Shichahai Historic and Cultural Area,

成占地6.8平方千米、融合金融要素市场和新兴科技园区的"新金融区"❸。这些以"新型""高端"产业为主导的园区建设如果付诸实践，将会带来怎样一轮新的旧城改造浪潮，令人不得不担忧。

2.2.4 北京旧城更新历程小结

总结北京自1990年代至今的旧城更新历程，经历了从房地产开发为主导的危旧房改造，到住房体制改革和城市现代化建设推动下的大规模拆迁和重建，再到新世纪在旧城人口疏解和整体保护思路的指导下，逐步转向小规模、渐进式的有机更新模式。总体特征如下：

（1）始终以政府作为更新主体，国家主导的特色显著。这也是我国各地城市更新的一个普遍特征。

（2）旧城更新的主题经历了住房改善—城市现代化—空间生产的过程。从早期旨在改善拥挤的住房条件、提升家庭居住质量，到第二阶段关注提升城市整体环境品质和生活便利性，再到新世纪全球化和消费主义浪潮背景下，以产业升级和品牌营造为主旨的社会空间更新。

（3）住房体制改革始终作为旧城更新的核心动力。回顾前两个阶段旧城改造的快速发展时期，无不借力于以住房制度为核心的体制改革，通过住房实物分配向货币化的转变，以及房改带危改等举措，引入市场和社会资金，激活住房市场。审视当前制约北京旧城保护与发展的住房产权格局混乱、改造资金不足、社会参与的制度性门槛等瓶颈问题，未来的突破口将依赖于住房体制改革的进一步深化。

（4）大事件成为旧城更新的催化剂。一个有意思的现象是，北京先后于1991年和1998年两次提出申办奥运会，正好对应两次旧城改造浪潮的兴起，带动大规模的固

Fujing Historic and Cultural Area, Liulichang Artistic Products Trading Center, Tianqiao Performance Area, Xidan Commercial Area, Dashilar Traditional Commercial Area, and Maliandao Tea Featured Commercial Area, which altogether accounted for 44.1% of the total land area of the District❷. Take the Financial Street (expansion plan) as example. The expansion plan, based on the original planning scope of 1.18 km^2, expanded the Financial Street westwards, southwards and eastwards to an area of 2.6 km^2, connecting southwards to Guang'an Industrial Park to form the New Financial Street of a larger area reaching 6.8 km^2 ❸. What will it bring to the urban regeneration of the Old City once those unconventional and high-grade industries are put into construction? There is no sign that we should be free from worries.

2.2.4 Summary

Since the 1990s, the urban regeneration of Beijing Old City experienced three distinct stages: dilapidated housing redevelopment driven by real estate development, large-scale demolition and urban renewal driven by housing system reform and urban modernization, and small-scale organic regeneration in the new millennium which is in line with the thoughts of population decentralization and overall preservation. It demonstrates the following features:

(1) Governments have played a leading role in urban regeneration all along, which is also a common feature of urban regeneration in China.

(2) The main theme of urban regeneration has undergone the process of housing improvement, city modernization, and space production. At the early stage, it was dedicated to easing the crowded housing conditions and improving the livelihood of households. The second stage focused on improving the overall quality of city environment as well as living convenience. In the new century, it shifted to socio-space regeneration aiming at industrial upgrading and regional brand building under the rising tide of globalization and consumerism.

(3) Housing system reform has always been the core driving force of Beijing Old City regeneration. The rapid pace of the first two stages relied on institutional reforms, in particular housing system reform. The measures, e.g. replacing the old housing allotment system with a new monetary and market-oriented one

图3 北京城市功能区分布示意图
（资料来源：刘佳燕）

Fig.3 Distribution of Beijing Functional Districts
(Source: Jiayan Liu)

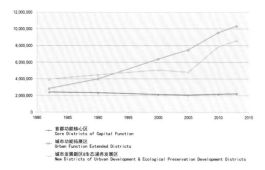

图4 1982—2013年北京市常住人口分布变化
（资料来源：数据来自北京市人口普查办公室，1982；庄宝国，1991；北京市第五次人口普查办公室，等，2002；庞江倩，2012）

Fig.4 Changes of permanent population distribution in Beijing during 1982–2013
(Source: data based on Census Office of Beijing, 1982; ZHUANG, 1991; The 5th Census Office of Beijing, et al., 2002; PANG, 2012)

定资产投资和城市现代化建设，推动旧城更新上升至全市乃至全国的重大战略地位。

2.3 旧城社会空间的转型特征

总结历经二十余载的更新改造历程，北京旧城社会空间主要呈现出以下四个方面的转型特征。

2.3.1 人口聚集效应日益显著，高密特征持续强化

1991年和2004年的两版北京城市总体规划中都明确提出了旧城常住人口向外疏解的目标。前者预期"从1990年的175万降至2000年的160万"；后者更是提出要从2004年的160多万减少到2020年的110万，意味着需要向外疏解50万人口，占原有人口的近1/3。但从实际情况看并不乐观。首都功能核心区常住人口自新世纪以来不降反升，从2005年的205万增长到2013年的221万（图3，图4）（邬春仙，2007；北京市统计局，等，2014）。伴随旧城居住用地的不断缩减，居住人口密度进一步加大，从1991年的5.41万人/平方千米增长到2008年的6.25万人/平方千米（钱笑，2010）。

大量人口向旧城聚集的效应更加突出地体现在日间活动方面。根据2005—2006年的普查数据，原东、西两城区内有夜间居住人口121万，而日间就业人口高达341万（景体华，2009），这还未加上进城开展旅游、购物、就医、就学、办事等活动的人群。数以百万计的日间活动群体涌入旧城，带来交通拥堵、公共资源供不应求等一系列问题。

分析旧城改造后人口不减反增的原因主要来自以下几

and advancing dilapidated housing redevelopment through housing reform helped to attract market funds and social capital, which fundamentally revitalized the housing market. Keeping that in mind, when we look at the current predicaments that restrain the Old City protection and development, e.g. complicated and confusing housing ownership pattern, shortage of regeneration funds, social participation issue and its institutional threshold, we may anticipate that future breakthroughs will largely rely on further deepening of the housing system reform.

(4) Major events acted as important catalysts to the urban regeneration of Beijing Old City. This is evidenced by an interesting phenomenon that, at the time of Beijing's bid for Olympic Games in 1991 and 1998, Beijing just initiated the first and second phases of the Old City regeneration; the events brought about large-scale fixed-asset investment and urban modernization construction, thereby elevating the process to a significant strategic status citywide or even nationwide.

2.3 Characteristics of socio-space transformation in Beijing Old City

A historical review over the urban regeneration of Beijing Old City in the past two decades indicates that its socio-space transformation presents four major characteristics.

2.3.1 Continuous population aggregation and increasing urban density

Both the Beijing City Master Plans of 1991 and 2004 put forward a clear objective to decentralize the permanent population in the Old City, with the former anticipating a decrease from 1,750,000 in 1990 to 1,600,000 by 2000 and the latter from around 1,600,000 in 2004 to 1,100,000 by 2020. However, the actual situation does not look optimistic. The number of permanent population in the "Core Districts of Capital Function", i.e. Dongcheng and Xicheng Districts, rebounded in the new century, from 2,050,000 in 2005 to 2,210,000 in 2013 (Fig. 3, Fig. 4) (WU, 2007; Beijing Municipal Bureau of Statistics, et al., 2014). Under the condition of decreasing residential land area, the residential density further increased from 54,100 people/km^2 in 1991 to 62,500 people/km^2 in 2008 (QIAN, 2010).

个方面：①受近年来土地成本快速上涨和开发商逐利的驱动，改造后建设量大幅提升，带来更多的居住人口；②用地置换后新增的办公和商务用地吸引更多的日间活动群体涌入旧城；③居民外迁难度日益增大，剩下的地块都是难啃的"硬骨头"。以往依赖大规模拆迁和人口外迁平衡旧城人口聚集效应的做法已经难以为继，新时期改造规模的大幅缩减使得旧城人口压力愈发凸显。

2.3.2 社会结构总体逐步提升，两极分化趋势凸显

20世纪末期以来，北京城市产业结构进入快速转型期，加上旧城改造后大量兴建的商品房带来高学历、高收入居民，推动旧城整体社会结构逐步提升。从受教育程度看，核心功能区内受高等教育人口占6岁及以上人口比重从1990年的14.2%，增长到2005年的30.0%，到2010年加速增长到39.8%，接近拥有大量高校和高科技企业的城市功能拓展区的水平（40.1%）（庄宝国，1991；邬春仙，2007；庞江倩，2012）。

但同时，社会群体的两极分化趋势也日趋显著。一方面，随着拆迁成本和地价的飞速上涨，旧城内更新改造项目日趋减少，并大多选择了对原有居民进行外迁安置的方式，越来越多的开发项目定位高端、奢华。以紧邻景山公园2014年开盘的某高档公寓楼盘为例，主力户型2居面积280～360平方米，3居459～560平方米，4居702～707平方米，总均价6000万～7000万元/套，单套最高售价1.40亿元（2014），客户群直指国际顶级富豪阶层。与之形成鲜明对比的是，大量从事商业服务业的外来低收入人口源源不断地向旧城聚集。以东城区为例，2010年常住人口的主要职业构成为商业和服务业人

Aggregation effects are reflected further in daytime activities. The census data (2005–2006) show that the daytime employed population in former Dongcheng and Xicheng Districts was up to 3,410,000 compared to the nighttime residential population of 1,210,000 (JING, 2009), and that is before counting those for tourism, shopping, medical services, education, and other businesses. Millions of people rushed into the Old City during daytime, leading to a series of problems like traffic jam and public resource shortage.

The phenomenon that the Old City witnessed a continuously increasing population after urban regeneration can be attributed to the following aspects. The surging redevelopment costs in recent years, together with the developers' profit-driven nature, greatly increased the floor area ratio, which consequently brought in more residents; After land-use replacement, the new offices and business areas attracted more population to the Old City for daytime activities; It becomes increasingly difficult for residents to move out and the remaining blocks are very hard to be redeveloped now. As the common practice relying on large-scale demolition and relocation to ease population aggregation is no longer effective, both the number and scale of urban regeneration projects decrease, which further aggravates the population pressure on the Old City.

2.3.2 Upgraded social structure and intensified social polarization

Since the 1990s, Beijing entered the stage of rapid industrial restructuring. The urban regeneration of the Old City resulted in the emergence of a number of new commercial apartments which attracted more highly-educated and high-income residents, promoting the upgrading of its social structure. In terms of education level, in the "Core Districts of Capital Function", the higher education ratio among those of over 6-year-old increased from 14.2% in 1990 to 30% in 2005 and the figure further climbed up to 39.8% in 2010, almost reaching the level of the "Urban Function Extended Districts" (40.1%) where a large number of universities and high-tech enterprises were located (ZHUANG, 1991; WU, 2007; PANG, 2012).

Nevertheless, social communities tended to be increasingly polarized. On the one hand, under the surge of demolition costs and land prices, the number of regeneration projects in the Old City

员（34.5%）、专业技术人员（27.2%）、办事人员和有关人员（24.7%）等。相对于高级人才主要集中在户籍人口，商业和服务业人员则在常住外来人口构成中占明显优势，比例高达64.3%，远远超出了北京市的平均水平（48.1%）（图5）（孙书振，等，2011）。

社会极化的背后是巨大的区域发展差距。2010年河北省来京人口为155.9万人，占北京常住外来人口的22.1%。2013年北京城镇居民人均可支配收入（40321元）相当于河北农村居民人均纯收入（9102元）的4.4倍，两者差距相比十年前进一步扩大（图6）。京津冀地区城镇群及城乡之间的发展不平衡，造成大量外来人口源源不断地涌入北京"淘金"。他们中的大部分寄身于那些在旧城改造浪潮中遗留下来的剩余空间，并进而由于生活环境的限制和公共服务的缺失，导致贫困继续传递影响下一代。

2.3.3 社会空间碎片化与隔离，产出效益差距显著

尽管北京旧城保护力度在不断加强，但经过多年的大规模拆建，历史保护街区已经支离破碎，城市景观亦日趋碎片化。在政府和地产商花重金打造的高科技产业园区和现代化商业街之间，穿插着一片片低矮拥挤的平房院落；在前门历史文化街区，全球奢华品牌旗舰店与廉价折扣店不协调地共存着（图7）。

另一方面是不同群体对有限空间资源的争夺和由此带来的空间隔离。各种"门禁社区"在胡同内不断涌现，从早期的单元大院到后来的高端商品房小区和豪宅大院，将原本就拥挤不堪的旧城空间进一步压缩肢解得所剩无几。即使在狭窄的街巷和胡同内部，私家停车位、绿植侵占公共步道的案例也屡屡发生（图8）。

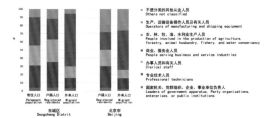

图5　2010年东城区和北京市常住人口主要职业构成比较
（资料来源：数据来自孙书振，等，2011）

Fig.5 A comparison of permanent residents' profession composition between Dongcheng District and Beijing City as a whole in 2010
(Source: data based on SUN, et al., 2011)

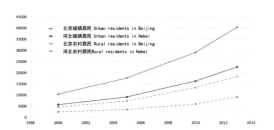

图6　北京市和河北省城乡居民人均年收入比较（单位：元）
（资料来源：数据来自中华人民共和国国家统计局，2014）

Fig.6 A comparison of per capita annual income of urban and rural residents in Beijing and Hebei (unit: RMB)
(Source: data based on National Bureau of Statistics of the People's Republic of China, 2014)

sharply decreased. Most of the projects put into effects preferred to targeting at high-end and luxurious apartments and relocating the original residents elsewhere, usually the urban fringe far away from the Old City. For instance, there is a high-grade residential apartment located in a walkable distance of around one kilometer to the north gate of Forbidden City. It was built up through urban regeneration and put onto the market in 2014. Its major house types include the two-bedroom apartment of 280-360 m^2, the three-bedroom of 459-560 m^2, and the four-bedroom of 702-707 m^2, Each apartment in 2014 had an average price of 60-70 million RMB, with the highest being 140 million RMB, targeting at the world's richest buyers. In strong contrast, a large number of migrants serving business and service industries with a low income constantly flow into the Old City. For example, in Dongcheng District in 2010, the permanent population was mainly composed of people serving commercial and service industries (34.5%), professional technicians (27.2%), and clerical staff (24.7%). Compared to advanced talents who are usually registered residents, people serving commercial and service industries accounted for a higher proportion (up to 64.3%) among the migrants in the Old City, far exceeding Beijing's average proportion of 48.1% (Fig. 5) (SUN, et al., 2011).

Behind social polarization is the huge disparity in regional development. In 2010 1,559,000 residents in Beijing were from the nearby Hebei province, accounting for 22.1% of the city's permanent migrants. In 2013, the disposable income per capita in urban Beijing was 40,321 RMB, which was 4.4 times of that of Hebei's rural residents (9,102 RMB), and the disparity was further enlarged compared to that of a decade ago (Fig. 6). Under the unbalanced urban-rural development in the Beijing-Tianjin-Hebei region, a large number of migrants constantly rushed to Beijing for what is called "gold washing" (the attempt to economic improvement). The majority of them live in the residual space left from the Old City regeneration waves. Due to their poor living environment and the absence of public services, their poverty has been, and will continue to be, passed onto the next generation.

2.3.3 Socio-space fragmentation and segregation resulting in significant disparity in output benefits

Despite its efforts of historic protection, Beijing Old City has lost its identity as most historic and cultural blocks were

碎片化的社会空间同时存在产出效益两极分化的问题。传统历史文化街区尽管在解决就业上发挥着重要作用，但长期以来受制于业态低端、种类单一、分布零散、收益低下等发展困境。以西城区为例，阜景、什刹海、大栅栏、琉璃厂、天桥等五大功能街区2010年的主营业务收入之和为140.9亿元，仅为德胜科技园的1/6，金融街的1/14（清华大学建筑学院，2014）（表1）。差距不仅来自各地区产业形式的不同，更重要的是产业背后支撑人群的差异。在天桥、什刹海等历史文化街区，传统文化遗产面临后继无人的严峻现实：原有居民受教育程度低，且老

fragmented during the large-scale demolition. Crowded low-rise houses are interspersed among high-tech industrial parks and modern commercial districts. Even in well-known historic and cultural district like Qianmen, there is inharmonious juxtaposition of luxury flagship stores and adjacent bargain stores (Fig. 7).

Along with spatial fragmentation is different communities' competition for limited spatial resources, which causes spatial segregation. "Gated communities" take various forms from danwei compounds to high-grade condominiums and grand courtyards. All these gated communities further compress and disintegrate the already crowded space of Beijing Old City. Even in narrow alleys and hutongs, it is not uncommon to see public footpaths illegally occupied, particularly by private cars and plants (Fig. 8).

Another problem with the Old City's socio-space is the

图7 前门大街的繁华景象和旁边平房区内的街巷空间形成强烈反差
（资料来源：刘佳燕）

Fig.7 Prosperity of Qianmen Street in contrast with its nearby alley space of the dilapidated traditional house area
(Source: Jiayan Liu)

表1 2010年西城区典型文化产业片区产业发展状况及与其他片区的对比
Table 1 Industrial development status of the typical cultural industry zones, in comparison with other areas

	法人单位数 Number of legal entities	资产总额（亿元） Total assets (100 million RMB)	主营业务收入（亿元） Major business revenue (100 million RMB)
阜景文化旅游街区 Fujing Cultural Tourism Area	276	113.5	22.3
什刹海历史文化旅游风景区 Shichahai Historic and Cultural Tourism Scenic Area	1533	162.1	79.3
大栅栏传统商业区 Dashilar Traditional Business Area	405	35.5	8.0
琉璃厂艺术品交易中心区 Liulichang Artwork Trading Zone	392	35.1	15.6
天桥演艺区 Tianqiao Performance Area	518	37.9	15.7
上述合计 Total (above)	3124	384.1	140.9
德胜科技园 Desheng Science Park	4980	6724.7	824.7
金融街 Financial Street	807	163000	1916.2

图8 私家绿植侵占公共步道
（资料来源：刘佳燕）

Fig.8 Footpath occupied by private green plants
(Source: Jiayan Liu)

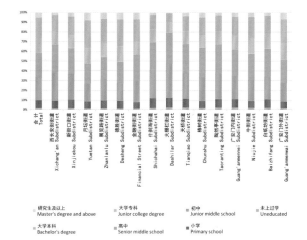

图9 2010年西城区各街道受教育人口比重
（资料来源：数据来自北京市西城区第六次全国人口普查领导小组办公室，等，2011）

Fig.9 Proportion of educated population in Xicheng District in 2010 (in unit of subdistrict)
(Source: data based on Xicheng District Head Office of the 6th Census, et al., 2011)

龄化现象严重，缺乏将传统文化和手工艺技巧进一步推广和市场化的能力；而更多的外来租户主要是低教育水平的服务业从业人员，对于传统文化的传承和发展在学习能力和意愿上都严重不足（图9）。上述问题将严重阻碍这类社会空间的可持续发展。

2.4 社区发展视角下的更新路径反思

由前文分析可见，北京旧城的更新历程体现出政府主导、体制革新、事件推动等自上而下的特色，但在整个主体性的建构中，社会和社区难觅身影。尤其当我们用社区发展的视角审视旧城更新及其带来的社会空间转型时，不得不对以下问题进行反思。

2.4.1 大规模空间生产不可持续，以住房体制改革推进发展权

1990年代以来，整个中国进入政府主导下空间生产的大舞台。旧城更新成为政府围绕"CBD""文化"等新时期热点概念重塑城市核心区新形象的重要途径。基于城市土地的全民所有，以及旧城内大量平房质量低下等原因，大规模推倒重建成为政府推进城市更新的普遍手段。它在很大程度上降低了城市改造的成本，创造了全球瞩目的高效率的城市转型，但同时也暴露出不可持续性等问题。

经历二十余年的改造历程，旧城内的"肥地"已经基本开发完，剩下的地块基本都是拆不动或赢利薄的"硬骨头"。旧城风貌保护限制日益严格，拆迁安置成本飞速上涨，正如访谈中一位居委会干部所说，"居民能走的都

unbalanced distribution of output benefits. While the historic and cultural blocks have played an important role in increasing employment opportunities, they have been for long constrained by some predicaments, e.g. low-end and highly homogeneous business, with low profits and scattered distribution. Take Xicheng District as example. Among the above-mentioned 10 major functional areas, the five areas of high historic and cultural values, that is Fujing, Shichahai, Dashilar, Liulichang and Tianqiao, generated a total revenue of only 14.09 billion RMB in their primary businesses in 2010, which equaled to 1/6 of that of Desheng Science Park and 1/14 of that of Financial Street (Table 1).(School of Architecture, Tsinghua University, 2014) The huge disparity is not due to the difference in industrial development modes, but more significantly to the difference in supporting groups engaging in the industries. In most historic and cultural areas (e.g. Tianqiao and Shichahai), cultural heritage does not seem to have qualified successors: locals are not properly educated and seriously aging, being incapable of promoting or marketizing traditional cultures and craft skills. Moreover, the places are attracting more migrants of low education levels who mostly work for service industries and possess a lower learning capability or willingness in promoting traditional culture (Fig. 9). These problems will seriously hinder the sustainable development of socio-space of the sort.

2.4 Reflection on urban regeneration method from perspective of community development

As argued above, the urban regeneration of Beijing Old City is characterized by an government-led, institutional innovation, event-driven and top-down mode, whereas most of the time the communities are neither visible nor active. Especially from the perspective of community development, when examining the whole process of urban regeneration, as well as the consequent socio-space transformation, the following problems should be acknowledged and considered.

2.4.1 Large-scale space production is hardly sustainable and development right should be highlighted in housing system reform

Starting from the 1990s, China entered a historical stage of

走了，留下来的要么是不愿走的，要么是搬不走的。"另一方面，政府主导、市场运作的模式提高了更新改造的进入门槛，将大量小规模的社会资本排斥在外。尤其是与旧城息息相关的当地居民长期被排斥在外：既无法作为"主体"主动参与更新改造，也无法作为"受体"享受更新改造后的升值效益。

以上问题带来了一种独特的绅士化现象，不同于传统西方绅士化现象中中产阶级迁入并推动环境改善的过程，这里体现为"空间的绅士化"在前，借此寄希望于推动社会结构的升级。而事实是，原有的中低收入居民或者被"置换出局"，其社会和空间区位被进一步边缘化，或者受限于产权的混乱，无法加入住房资本的自我循环和升值的游戏，被长久固化在破败的平房区和社会底层。

西方城市更新经过对过去半个多世纪历程的反思，正逐步将促进广泛而公正的公共参与作为核心议题。以实现平等的发展权为目标，参与不仅仅体现在"知情"和"发声"的层面，更应从根本上赋予居民参与旧城更新的权利和能力。如今，住在平房区的居民大部分可以说是"抱着金碗要饭吃"：他们的住房只有进入更新程序才拥有可兑现的市场价值；然而能否进入更新程序，何时进入，以及如何进入，都无法由他们说了算。要改变当前旧城更新的困境局面，一个重要的突破口是探索更加灵活多样的住房产权置换、交易、金融化等形式和市场准入机制，以深化住房体制改革保障和推进旧城居民的发展权。

2.4.2 旧城社区日趋杂化，急需重建社区生活共同体

随着旧城改造的推进和单位制的解体，多年来稳定而宁静的旧城社区日趋杂化，居住人口的构成逐步多样化，

government-led space production. Old city renewal became an important way for governments to reshape the image of the city center centering on the hot concepts of "CBD" and "culture". For the state ownership of urban land, together with massive low-rise housing with poor quality in the old city, large-scale demolition and reconstruction was a common approach for governments to promote urban renewal. This, to a large extent, lowered the costs of urban redevelopment and created a highly-efficient urban transformation model. Having attracted a large volume of global attention, this model meanwhile exposed increasing problems which proved its non-sustainability.

After 20 years' transformation, valuable land in the Old City has been almost developed, and the rest is largely undemolishable or unprofitable. With stronger restrictions on old city preservation and fast rising costs of demolition and relocation, "all movable residents have moved, except those who would not or could not", as one community residents' committee member remarked. On the other hand, the government-led market-promotion mode elevated the threshold of such regeneration projects, which excluded many small-scale social capitals. Local residents had been particularly excluded throughout the process: neither could they participate in the regeneration as "subjects", nor did they benefit from land value appreciation as "beneficiaries".

All the above led to a distinctive gentrification process, i.e., spatial gentrification precedes, expecting to promote the upgrading of social structure, in contrast to the traditional gentrification process in western counties whereby middle classes moving in and then promoting environmental improvement. The fact is that original residents with middle and low incomes were displaced as their socio-spatial positions became further marginalized. Or in other times, as complicated and unclear property rights discouraged their participation in the self-circulation and appreciation of housing capital, they were permanently "locked" in the dilapidated low-rise courtyards and lower social class.

Urban regeneration in Western cities has gradually regarded extensive and equal public participation as a core theme. Aiming at equal development rights, participation is not only reflected on the level of "knowing" and "voicing", but more importantly, on the empowerment of the local residents in the whole process of regeneration. Nowadays, most residents in low-rise courtyards are

流动性日益显著。

在旧城改造的早期，也就是本文提到的前两个阶段，改造后的居民以回迁为主，加上通过购置商品房新入住的，主要都是因为就业或生活便利的原因选择入住旧城，他们在当地居住和消费，全面融入当地的社区生活。

而近十年以来，随着房价高企，改造回迁的比例越来越少，三类新群体成为迁入旧城的主力军：①高端置业者。他们看中的是旧城的升值潜力（作为投资品）和"符号消费"的生活方式（作为家庭度假偶尔光顾的第二居所，或宴请宾客的会所），他们在当地鲜有消费，更毋论地方性的社会交往。②"基础教育移民"。北京东西城的基础教育长期以来拥有无可比拟的优势，每年大量的学龄儿童家庭为了争取入学资格而涌入旧城。举一个笔者2013年在北京旧城调研中发现的典型案例：三口之家为了让孩子获得西城区一所市重点小学的入学资格，花费180万元购置了一间（带入学指标的）9平方米的平房，以每月1500元出租给外来务工者，然后每月花6000元租赁学校旁边的一套两居室，母亲带着孩子住在里面，周末时间一家人再回到郊区的大房子团聚。③低端服务业群体。旧城发达的商业、办公和旅游产业吸纳了大量的服务业人口，他们通常以合租的方式几人甚至几十人租住在大杂院或简易楼房中，每天早出晚归，几乎没有周末或假期。

这三类新居民在日常的社区公共生活中几乎不见身影，与邻里鲜有交往，特别是第一和第三类群体，居委会描述"基本当他们不存在，平时也见不到人，顶多消防安全检查的时候打个招呼"。他们与旧城的关系纽带全部围绕房子展开，但又仅限于房子——从交换价值（市场价格和入学资格）到栖身之所——而依托住房的社区交往和情

like "begging for bread with a gold bowl": their houses can only be redeemed with market values after entering the regeneration procedure, but they themselves can hardly decide whether, when, or how this will happen. Hence, to break the current predicament, an important breakthrough would be to explore more flexible and diversified forms of housing property replacement, trade and financialization, as well as the market access mechanism, so as to further deepen the housing system reform and to safeguard residents' development rights in the Old City.

2.4.2 Increasing diversification of Old City neighborhoods highlighting community rebuilding

Under the urban regeneration and the collapse of danwei system, the originally stable and peaceful neighborhoods in Beijing Old City were increasingly transformed. There was a more mixed population, as well as a stronger mobility.

At the early stage of Old City regeneration (the first two stages as introduced in part two of this paper), the majority of the original residents would be resettled back afterwards. Even those moving in by purchasing commercial houses, they would reside in the Old City for employment and living conveniences, fully engaged in local community life.

However, accompanied with the boom in housing price over the past decade, the number of moving-back residents was decreasing. The Old City's new residents mainly consist of the following groups: high-end house buyers. They prefer the Old City's appreciation potential (house as a commodity), as well as the life style of "symbol consumption" (the second home for occasional family tourism, or a venue to receive guests). They consume little at local markets, not to mention their contact with local society. Basic education migrants. As Dongcheng and Xicheng Districts possess supreme advantages in basic education, a large number of households rush to the Old City's real estate market every year to qualify their children for local school admission. According to a survey we conducted in 2013, a 9 m^2 crude shelter was sold at the price of 1.8 million RMB for its quota of admission qualification of a nearby excellent primary school. The buyer leased it to migrant workers at a price of 1,500 RMB per month, then rented a two-bedroom apartment at a price of 6,000 RMB near the school, so that the mother could

感联系在这里难觅踪影。曾经高度紧密的旧城社会网络如今日益呈现出稀疏化甚至解体的趋势，如何重建社区生活共同体成为旧城社区规划和社会建设中不可忽视的一个重大挑战。

2.4.3 关注"不可见"的弱势群体，重视资本为本的能力再造

在旧城更新的浪潮中出现了一批弱势群体，亟待关注。一方面，他们是更新项目中几乎"不可见"的群体，既非政府产业结构升级计划中希望引入的"三高"（高学历、高职位、高收入）人才，亦非地产开发项目中潜在的中高收入阶层。另一方面，他们唯一可见的时候是出现在拆迁改造和向外疏解的名单上，包括原有居民、租房户、服务业从业人员等。可以说，旧城更新离他们很近，他们的家、邻居、每天走过的街道可能随时面临改造；但实际上更新又距离他们很远，或者多年翘首以盼拆迁通知的早日降临，或者在某个清晨突然获悉自己的住房或店铺即将被拆除，他们始终是被动地接受着周遭的改变。

随着近年来政府对于公众参与的逐步重视，在一些街区保护规划和社区发展规划中，基层政府干部、专家和技术人员通过问卷调查、访谈等形式走进社区，尽可能了解居民需求并反馈到规划的制定和实施中。但这种方式往往因为现实中过多的问题和需求超出了政府的有限能力，而最终难有作为。更重要的是，长期过度依赖外部力量容易导致社区自身"造血能力"的丧失。因此，社区发展的模式需要从"需求为本"转向"资产为本"，从单纯强调外部力量的投入转向重视社区自身资本和能力的建设。社区的每一个居民，不仅是服务的享用者，同时也是社

accompany her child during weekdays, while at weekends enjoying their family life in the spacious suburban villa. low-end service industry employees. The Old City has attracted a large number of service industry employees for its developed business, office and tourist industries. These people would share a courtyard or a simple house with up to a dozen of other people; they work from dawn until dark, without weekends or holidays.

The three groups of new residents in the Old City are almost absent from community life and they have little contact with neighbors. Especially, the community residents' committee would, as one committee director interviewed said, "take them (the above-mentioned especially the first and third groups) as invisible beings because we can hardly see them. We might have a word only during the fire safety inspection". Their tie with the Old City seems to center on houses yet restricted within houses: from its exchange value (market price or school entry eligibility) to a place of living, whereas they have little contact and emotional connection with local neighborhoods. The once highly close social network in the Old City is increasingly disintegrating, and is now looser than ever. Consequently, how to rebuild solidary communities should be a major challenge during current community planning and society building in Beijing Old City.

2.4.3 Paying attention to the "invisible" disadvantaged group, emphasizing capital-based capacity reconstruction

During the Old City regeneration, there are a group of disadvantaged people who are needed to be highly regarded. They are mostly invisible, belonging to neither the "three high" (high education, high professional title, and high income) population that the government would like to attract during the industrial structure upgrading, nor the potential middle-or-high-income classes that could afford housing properties. The only time when they are "visible" is on the name list for demolition. These people include low-income locals, tenants, and small-scale service industry employees, etc. While the Old City regeneration is physically near them - their houses, neighbors and living streets could be regenerated at any time, yet they could only passively accept the changes, whether expecting them or not.

As the governments pay increasing attention to public

力量的创造者和传播者,这就需要在旧城更新规划中充分发掘和培育社区的物质资本、文化资本和社会资本。例如,东城区史家社区在2013年建立了北京首个"胡同博物馆",展示的不仅有社区居民主动捐赠的老照片和老物件,还包括"老北京叫卖"等非物质文化遗产,使得博物馆在与社区的互动中体现出非凡的特色和生机。

2.5 结语

北京旧城不仅是历史名城的核心物质载体,更是古都文化传承和复兴的社会空间载体。旧城整体保护的实现,不仅体现在总体风貌和空间格局的保护和延续,更需要将旧城视为一个完整社区,致力于促进其社会空间的协调发展和人文活力的持续再生。如何结合住房体制改革推进旧城居民全面参与旧城的更新发展,如何跳出传统保护规划单纯"保空间"的局限,重视地方资本和精神的培育,推动社区生活共同体的重建和社区自我发展能力的再造,这些都是当前旧城更新和社会空间转型过程中不可忽视的重大挑战。

participation, residents' demands and voice have been made known and incorporated into the regeneration. Government departments, experts and planners step into communities and conduct surveys or interviews when they formulate preservation plans or community development plans. However, it usually contributes little because of the irreconcilable contradiction between community's endless demands and government's limited capability. More importantly, for their over-reliance on external resources, communities would lose their function of self-development. Therefore, community should shift its development mode from "demand-oriented" to "capital-oriented", from heavy reliance on external support to its own ability enhancement. It requires to discover and improve the community' material capital, cultural capital and social capital. For example, Shijia community in Dongcheng District established Beijing's first "hutong museum" in 2013, not only exhibiting local residents' donations of old photos and collections full of common memories, but also providing a platform for the inheritance and development of non-physical cultural heritages, like vendors ' hawking, which renders the museum one of the most vigorous and featured scenery in the Old City.

2.5 Conclusions

Being the core physical center of the historic city, Beijing Old City is in the meantime the carrier of socio-space for the ancient capital's cultural inheritance and renaissance. It means that the overall protection of the Old City should not only entail the protection and continuation of its overall features and spatial pattern, but also promote the coordinated development of socio-space and the sustained revitalization of cultural vitality, the latter of which requires us to view the Old City as a complete community of sustainability. As such, Beijing Old City regeneration and socio-space transformation will face several major challenges: how to empower residents' all-round participation in the urban regeneration through housing system reform, how to get over the space-only restrictions of conventional protective plans while cultivating and revitalizing local capitals and spirits, as well as how to facilitate the rebuilding of a community and its self-development capacity, etc.

参考文献 / References

北京市第五次人口普查办公室，北京市统计局，2002. 北京市2000年人口普查资料[G]. 北京：中国统计出版社.

北京市人口普查办公室，1982. 北京市第三次人口普查手工汇总资料汇编[G]. 北京：北京市人口普查办公室.

北京市统计局，国家统计局北京调查总队，2009. 北京六十年（1949—2009）[M]. 北京：中国统计出版社.

北京市统计局，国家统计局北京调查总队. 北京市2013年国民经济和社会发展统计公报［EB/OL］. （2015-02-12) [2015-03-31]. http://www.bjstats.gov.cn/tjsj/tjgb/ndgb/201511/t20151124_327764.html.

北京市统计局，国家统计局北京调查总队，2014. 北京统计年鉴2014[M]. 北京：中国统计出版社.

北京市西城区第六次全国人口普查领导小组办公室，北京市西城区统计局，北京市西城区经济社会调查队，2011. 北京市西城区2010年人口普查资料[G]. 北京：北京市西城区统计局.

北京市住房和城乡建设委员会，2008. 2008北京建设年鉴[J]. 北京：北京市住房和城乡建设委员会.

北京市住房和城乡建设委员会，2010. 2010北京建设年鉴[J]. 北京：北京市住房和城乡建设委员会.

董光器，1998. 北京规划战略思考[M]. 北京：建筑工业出版社.

方可，2000. 当代北京旧城更新[M]. 北京：中国建筑出版社.

冯熙，2014. 关于历史文化街区保护的若干思考[J]. 东城人大，（2）：19-24，34.

景体华. 西城区人口动态分析与疏解对策研究［EB/OL］. （2009-08-27）［2015-03-31］. http://rkjshw.bjxch.gov.cn/RKJSxxxq/536262.html.

庞江倩，2012. 北京市2010年人口普查资料[G]. 北京：中国统计出版社.

钱笑，2010. 北京居住空间的发展与变迁（1912—2008）[D]. 北京：清华大学.

清华大学建筑学院. 2014. 西城区人口、资源、环境可持续发展研究报告[R].

孙书振，杨舸，郭永，等. 东城区的人口调控方式的研究——从产业结构出发[R]//第六次全国人口普查办公室，2011. 东城区第六次全国人口普查课题研究报告. 北京：第六次全国人口普查办公室.

邰磊，2010. 以南池子改造为例浅谈北京四合院的保护[J]. 青岛理工大学学报，（5）：32-36.

魏科，2005. 1990：北京两次大规模危改[J]. 北京规划建设，（6）：71-76.

邬春仙，2007. 2005年北京1%人口抽样调查资料[G]. 北京：中国统计出版社.

郑珺. 近十年来北京市的文物保护工作［EB/OL］. （2010-02-01）［2015-03-31］. http://dangshi.people.com.cn/GB/138903/138911/10901625.html.

政协北京市委员会城建委，等. 关于报送政协北京市第九届委员会常务委员会《关于北京城区危旧房改造问题的建议案》的函（京协厅函〔2000〕119号）［EB/OL］.［2015-03-31］. http://www.mohurd.gov.cn/zcfg/xgbwgz/200611/t20061101_159557.html.

中华人民共和国国家统计局，2014. 中国统计年鉴2014[J]. 北京：中国统计出版社.

周乐，2002. 对北京当前大规模危旧房改造的思考[J]. 北京规划建设，（4）：43-47.

朱自煊，2001. 专家纵论北京危旧房改造与古都风貌保护——北京城科会"历史文化名城与危旧房改造研讨会"回顾[J]. 北京规划建设，（4）：13-17.

庄宝国，1991. 北京市第四次人口普查手工汇总资料[G]. 北京：中国统计出版社.

风格主义新城镇
MANNERIST NEW TOWNS

盖里·海克
宾夕法尼亚大学设计学院荣退教授
清华大学建筑学院客座教授

Gary Hack
Emeritus Professor, School of Design, University of Pennsylvania
Guest Professor, School of Architecture, Tsinghua University

中文翻译：梁思思，刘健
Chinese translation by Sisi Liang and Jian Liu

3.1 引言

至少在罗伯特·文丘里、丹尼斯·斯科特·布朗和史蒂夫·依泽诺1972年出版《向拉斯维加斯学习》之后，风格主义已然成为世界城市设计的主流。风格主义一词在建筑领域沿用已久，上可追溯至16世纪和17世纪的风格主义运动，夸张的比例、压抑的空间、拥挤的画面，由此形成强调形式高于功能的习惯。关于风格主义，还有另外一种被更加普遍接受的诠释，即它是从别处引入新的观念，在无视历史、目的与功能的情况下，将其融入当地文脉，进而"参照某某样式"创造新事物的过程。令我感到惊讶的是，在中国，有许多新开发的的商业建筑，不论其结果好坏，都承袭了这种观念。尤其在上海出现的几座具有明显主题的新城镇，促使我对风格主义城市设计引发的问题进行了思考。

作为上海"1966"城镇体系规划的组成部分，"一城九镇"被提议成为区域发展的增长极；其中至少有6~7座新城镇借用了海外城市和国家的标志特色，作为其城市建设的主题。可以说，这是典型的风格主义城市的建造方式，引进其他地方的理念，将其植入当地城市发展之中。

松江新城，常被称为泰晤士小镇，是其中最被浓墨重彩描绘的新城镇；其次是自诩为汽车制造业首都的安亭新镇，由阿尔伯特·施佩尔及其合伙人事务所设计的德国风情小镇；然后是浦江新镇，由格里高蒂国际联合公司和其他意大利建筑师设计而成。它们共同代表了城市设计中的欧洲趣味，其灵感分别来源于传统的英格兰村庄、德国社会住房和意大利未来主义思潮（图1~图3）。

中国的新城镇建设如此直白地套用外国理念，这背后

图1 泰晤士小镇
（资料来源：Gray Hack）

Fig.1 Thames Town
(Source: Gray Hack)

3.1 Introduction

Mannerism has been a central point of contention in the world of urban design, at least since the publication by Robert Venturi, Denise Scott Brown and Steven Izenour of *Learning from Las Vegas* (1972). In the world of architecture, the term mannerism has a much longer history, dating to the stylistic movement in the 16th and 17th Centuries that exaggerated proportions, compressed space, crowded the picture frame and adopted an affected habit that overemphasized style over functions. There is also a more commonsense meaning to mannerism, which includes the process of importing ideas from other places, and incorporating them in the context at hand without great regard for history, purpose or function – creating things "in the manner of…" It strikes me that much of the new architecture of commercial development in China, for better or worse, has adopted this attitude. I have been particularly stimulated to consider the issues raised by mannerist urban design by the emergence of several explicitly themed new town areas in Shanghai.

As part of the 1-9-6-6 plan for Shanghai, a ring of 9 new cities were proposed as growth poles for regional development. At least 6 of them, perhaps 7, involve borrowing iconic ideas from cities and countries abroad, to serve as a theme for the new city center that's being created. So this is really a mannerist city that's being created, that imports ideas from other places and layers it onto the urban development.

Songjiang, which is often called Thames Town, is the new town that's most often portrayed. Anting New Town, a German-inspired town designed by Albert Speer & Partners in the area that bills itself as the capital of the auto industry, is a second example and a third is Pujiang New Town, designed by Gregotti Associates and other Italian architects. Together they represent three European tendencies in urban design–inspired by traditional English villages, German social housing and Italian futurism (Fig. 1–Fig. 3).

的逻辑是什么？原因有很多，其中最重要的，这是一种快速建立城市品牌、提升市场吸引力、塑造独特内在个性的方法。在20世纪的消费文化里，最强有力的理念便是品牌化，通过关联知名的理念来推销商品和经验。生活方式的品牌化是一种人为的理念，因为生活方式通常需要很长时间的发展才能形成；城市的品牌化则意味着人们可以创造出一种快捷的生活方式，或者创造出对代表了某种特定生活方式的场所的快速认知。

这些风格主义新城镇背后的另一个原因是环境决定论，认为生活模式与当地经济的相互呼应将有利于新城人口的就业。这种观点在美国曾被称为"福特主义"，因为亨利·福特坚持他在底特律的雇员要居住在模式化住区中而得名。有趣的是，这竟是开发建设安亭新镇的动因之一。安亭新镇位于大众汽车大型生产设施和通用汽车合资工厂所在地，周边还分布着汽车设计中心以及许多上海地区的大型汽车经销商，由此推断新城应从德国，尤其是德国类似的汽车制造发达地区汲取灵感，显然是符合逻辑的，或者至少不是离谱的想法；同样也不难判断，新城居民都将从事汽车的生产与设计，以及管理国际汽车公司的复杂生产流程。于是，诸如沃尔夫斯堡这样的城市就会进入人们的脑海，它在历史上曾经是大众汽车的全球经营和生产总部。

风格主义新城镇背后的第三个原因，似乎是期望通过委托来自特定国家的设计师和规划师进行新城的规划设计，能够同时吸引这些国家投资到新城的建设和市场推广中。浦江新镇早期的部分投资商就是意大利的保险公司，瑞典投资商则在斯堪的纳维亚主题的罗店新镇建设中占有巨大份额。

What is the logic of expropriating ideas so clearly from foreign sources and using them for new cities in China? The reasons are many, but foremost it's a way of instantly branding the cities and increasing their market appeal. It's a way of creating a distinct and coherent identity. The most powerful idea in 20th century consumer culture is the idea of branding, promoting goods and experiences through association with known ideals. Lifestyle branding is an artificial idea, because lifestyles typically develop over long periods of time. The idea of branding cities suggests that you can create an instant lifestyle, or an instant identification of a place with a particular lifestyle.

A second idea that I believe is behind these mannerist cities is environmental determinism: that the new occupations of people in the new cities will benefit from living patterns that are in concert with the economies of places. This notion was once called "Fordism" in the US, after Henry Ford's insistence that his employees in Detroit be housed in model settlements. Interestingly, this is one of the motivations for Anting New Town, located in the area of Volkswagen's large production facility and General Motors' joint venture plant. Also located there are the design centers for automobiles, and many of the largest automotive dealers in the Shanghai region. So it would seem logical, or at least not a big leap, to conclude that one should look to Germany for ideas about a new town, particularly the kinds of places where autos are made in Germany. The inhabitants, one might conclude, might be similar: people manufacturing cars, designing them, and managing the complex processes of running an international automobile company. Towns like Wolfsburg come to mind, historically the headquarters of Volkswagen's worldwide operations and production.

The third motivations for this strategy seems is that by hiring designers and planners from particular countries, they might also then be able to attract investment from those countries in building and marketing the new town. Some of the early investors in Pujiang were Italian insurance companies, and Swedish investors played a large role in establishing the Scandinavian- themed new town of Luodian.

图2 安亭新镇
（资料来源：Gray Hack）

Fig.2 Anting New Town
(Source: Gray Hack)

图3 浦江新镇
（资料来源：Gray Hack）

Fig.3 Pujiang New Town
(Source: Gray Hack)

最后，快速完成新城建设的需要可能也是一个原因。一方面希望快速完成新城建设，一方面又希望新城有别于普通的城市扩张，在这种情况下，选择另外一种形式的聚落作为模板就成为十分有效的方法，模仿知名的城市和城镇便可以达到这个效果。

所以，在上海新城镇建设的背后至少有上述四个方面的原因。那么结果如何呢？是否形成了宜居的社区，超越了传统的城市开发？对这种方式持批评意见的人们往往将这些新城镇视为迪士尼乐园，认为它们不过是从海外大众消费文化中吸取想法，继而转交给中国消费者，一如迪士尼在许多国家的做法；他们认为这只是娱乐，而非严肃的城镇建设。

3.2 迪士尼乐园及其理念输出的案例

迪士尼乐园本身就是多个场所的拼凑与杂烩。它的标志性主干大街是美国诸多主要街道的混合，带有来自不同历史时期和不同地方的特点，巧妙地利用了人们的怀旧之情。在美国，这条主干大街之所以如此受欢迎，原因之一就是这样的街道正在快速消失，人们渴望见到拥有许多本土商店、二楼设有牙医诊所的老城镇，而迪士尼乐园则以3/4的尺度重现了这些景致。这意味着，根据风格主义的做派，可以在较小的空间范围内压缩容纳更多的活动。

迪士尼乐园的主干大街最早出现在加利福尼亚州，然后被复制到佛罗里达州的迪士尼世界，现在在香港、东京、巴黎和其他地方都能看到类似的迪士尼主干大街。虽然它们会有些许细微变化，例如东京的迪士尼主干大街覆有雨棚以适应当地频繁的降雨，但是世界各地的迪士尼乐

Finally, the necessity to build the towns quickly may have played a role. If one wishes to differentiate them from ordinary urban expansion, but build rapidly, it is useful to have a template to follow for another type of settlement. Emulating known towns and cities can provide this.

So at least those four ideas played a role in in the formation of Shanghai's new towns. How good are results? Have they resulted in communities that are livable, and a step above conventional development? People who are critical of the approach taken often dismiss the new towns as a form of Disneyland. They see them as importing ideas drawn from mass consumer culture abroad and offering them to Chinese consumers, much as Disney has done in several countries. It's entertainment, they say, not serious town building.

3.2 Disneyland and Transporting Examples

Disneyland, itself is a pastiche of many places. Its iconic Main Street is an amalgamation of many main streets in the United States, with characteristics drawn from different periods of history and different places. It plays on nostalgia. One of the reasons Main Street has been so attractive in the United States is that main streets are fast disappearing. People crave the old town with a variety of local shops and the dentist's office on the second floor. Disneyland reproduced them at about three-quarters scale, so that in the spirit of mannerism, more activities could be compressed into a smaller space.

Originated in California, then reproduced in Florida in Disney World, you can now find similar Main Streets in Hong Kong, Tokyo, Paris and other places. There are subtle changes – Main Street in Tokyo is covered by a roof to cope with the frequent rainfall, as an example – but Disneylands worldwide convey similar themes, and are part of a cross marketing program with Disney films, games and merchandise. So the question is, are these new towns that had been created just another variation of Disneyland? Are they part of a process of developing a universal culture of city building? (Fig. 4)

园都传达着同样的主题，是由迪士尼的电影、游戏和商品共同组成的市场营销的组成部分。那么问题是，上海的这些新城镇真的是迪士尼乐园的另一种变体吗？它们是否成为城市建设全球文化发展过程的组成部分（图4）？

当然，迪士尼并非首个借鉴这种成功的企业。在美国过去40年建设的新镇里，影响最为深远的新镇是锡赛德镇，它也绝对是同等规模的城镇中，被描述最多、被访问最多的城镇。锡赛德镇是围绕一个小型城镇公共绿地建成的滨海度假城镇，公共绿地周围是一圈店铺和独立住宅，每栋建筑都有相同的特点，但每栋住宅又都与众不同。它们遵循一套非常严格的规范和导则，决定了屋顶坡度、建筑材料、开窗形式和房屋颜色等内容，同时市镇规则也决定了什么可以建、什么不可以建。

锡赛德的城镇规划也同样重要：布局了多条装饰有露台的环形道路，位于镇中心的表演场地以及划定院落边界、将私人空间与公共领域区分开来的围墙。这种平面布局的缘起不在佛罗里达，而是新英格兰地区，更确切地说，是马萨诸塞州玛莎葡萄园里卫理公会教徒野营聚集场所。19世纪初期，许多家庭开始在室外围绕亭榭支起帐篷避暑，他们在那里举行集会、宗教仪式和教育讲座；之后，在一小块一小块的土地上，出现了如同玩偶小屋一样的木屋，逐渐取代了原有的帐篷。锡赛德的设计师们在寻找设计原型时，注意到了这个野营聚集场所。因此，锡赛德模式没有任何本土化内容，它只是吸取了麻省其他一些避暑胜地的精华，并在其中加入了少许欧洲村庄规划的理念（图5）。

同样在英国，在建设庞德伯里镇时，查尔斯王子和他的设计师里昂·克里尔汇集了许多英国和欧洲城镇的优秀创意。最终的结果是，这个小镇看起来颇具爱德华式风

Disney was, of course, not the first entrepreneur to borrow from beloved examples. One of the most influential new towns built in the United States in the last 40 years is Seaside, certainly the most written about, and the most visited place of its size. Seaside is a resort town built around a little town common, near the ocean. It has series of shops ringing the common and individual houses, each designed with similar characteristics. Every house is unique, but they abide with a very strict set of rules and guidelines that determine the slope pitched roofs, building materials, the types of windows, and colors of houses. A town code determines what you can and cannot build.

The town plan of Seaside is equally important with circles furnished with gazebos, places for performances at the town center, and individual fences that define yards and differentiate private spaces from the public realm. Its origins were not from Florida but from New England; specifically, the Methodist Camp Meeting Ground on Martha's Vineyard in Massachusetts. At the beginning of the 19th century, families began spending summers in tents surrounding an outdoor pavilion, where they gathered for meetings, religious services and educational speeches. In time, cottages, on very small lots, almost like dollhouses, replaced the tents. In looking for a pattern, the designer of Seaside focused on the Camp Meeting Ground. There was nothing indigenous about Seaside's pattern; it combined the best of other summer resorts in Massachusetts and added a dash of European village planning (Fig. 5).

Similarly in creating Poundbury in England, Prince Charles and his designer, Leon Krier, created an amalgam of good ideas from many British and European towns. The result is a town that looks Edwardian, but unlike any other town, since it was created instantly not through a lengthy process of each new generation adding a building or two (Fig. 6).

图4 迪士尼乐园
（资料来源：Wikipedia CC）

Fig.4 Disneyland
(Source: Wikipedia CC)

图5 锡赛德镇
（资料来源：Gray Hack）

Fig.5 Seaside
(Source: Gray Hack)

格，但却又与其他城镇明显不同，因为它是短期内快速建设的产物，而没有经过迭代新增的漫长过程（图6）。

美国有许多类似的开发项目。1909年初，纽约市皇后区的郊外大部分还是农田，格罗斯夫纳·阿特伯里设计了一座名为森林山的新村，恰巧位于一条新建地铁线的上方。他的设计灵感来自花园城市运动中建成的英国小镇，以及他所向往的具有场所品质的传统英格兰小镇。很快，这里的网球俱乐部成为美国网球公开赛的发源地，森林山村因此名声大振。从设计草图可以看出，这是一个颇具浪漫情怀的小镇，中心是一座酒店和若干商店，周围环绕着独立或双拼住宅以及公寓。时至今日，这里仍是纽约市最好的住区之一，并且与周边地区相比，升值更快，人们宁愿等待数年在此购房。这似乎是上海新城镇之前的先例（图7，图8）。

3.3 松江——泰晤士小镇

泰晤士小镇是更大规模的松江新城的核心板块，依托上海地铁9号线的终点站——松江新城站建设而成。车站周围环绕布置着各类机构的大型综合体和高密度住宅，相信未来还会有更多新的建设不断出现。一个精心维护、郁郁葱葱的带状公园将地铁车站与松江新城的象征性中心——泰晤士小镇联系在一起。遗憾的是，新城里几乎没有公共交通，大多数人不得不依靠私家车上下班或往返地铁车站。泰晤士小镇的中心建在一处可以俯瞰人工湖面的岬角上，其中分布着公共和私人的办公机构、一个艺术中心，以及一些商店、餐馆和住宅。穿过政府办公楼前新月形的拱道，是一系列尺度宜人的街道，基本是德文郡和其

In the United States, there are many such projects. Beginning of 1909, in a suburb of Queens, New York City that was mostly farmland at that time, Grosvenor Atterbury designed a new village called Forest Hills, which was on a newly created subway line. His sources were English towns that were being built as part of the Garden City Movement, but also traditional towns in England with the kind of qualities of places that he sought to create. It quickly gained through its tennis club as the original location of the U.S. Open. As the sketches show, it was a romantic village, with a hotel and shops at the center, and archetypal one and two family houses and apartments surrounding it. It remains today one of the best residential areas in New York City. It has gained value faster than other nearby areas, and people wait for years to buy a house in this area. This seems an obvious precedent for Shanghai's new towns (Fig. 7, Fig. 8).

3.3 Songjiang–Thames Town

Thames Town is centerpiece of a much larger new town, Songjiang, located at the terminus of Shanghai's No 9 Metro line. Large institutional complexes and dense housing, with more to come in the future, surround the main metro stop, Songjiang Xincheng. A lushly planted and well-maintained linear park connects the station area to the symbolic center of Songjiang, Thames Town. Unfortunately, there is minimal transit in the town, and most people are forced to rely on automobiles for their journey to work, or to the transit station. In Thames Town, a new town center, with public and private offices, an art center, shops and restaurants, and housing has been built on a promontory overlooking a manmade lake. Through the archways of the crescent shaped government complex are a series of pedestrian scaled streets modeled on Devon and other English country towns (Fig. 9).

图6 庞德伯里镇
（资料来源：Lan Skelley）

Fig.6 Poundbury
(Source: Lan Skelley)

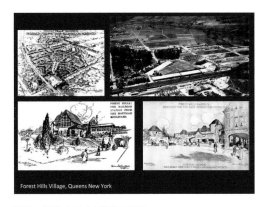

图7 纽约森林山村设计草图
（资料来源：Grover Atterbury）

Fig.7 Sketches of Forest Hills Village, New York
(Source: Grover Atterbury)

他英国乡村小镇的翻版（图9）。

与大多数新开发区不同，泰晤士小镇看起来人声鼎沸。人们在街上漫步，在餐馆进餐，在商店购物；这些商店数量不多，应该是熬过了开业之初的艰难，坚持到现在。最为突出的是，这里是拍摄婚纱照的最佳之选；天气晴好时，会有数十对夫妇穿着租来的服装，在街道里，在教堂前的草地上，在湖边和运河沿岸，以及泰晤士小镇里一切风景如画的角落，摆出各种姿势拍照留念。这里如同舞台布景般的别致景色吸引了许多婚庆摄影企业，在小镇的主要街道上开设了多家店铺（图10，图11）。

是什么使泰晤士小镇如此具有吸引力？尺度是原因之一，这里的建筑尺度宜人，与太多新区开发里裙房之上高塔林立的情况形成鲜明对比；功能混合是原因之二，公共建筑、教堂、住房和商店鳞次栉比；空间肌理是原因之三，小型的绿色空间、广场、廊道和私人绿坪共同组成了连续的开敞空间。社区的平面布局围绕一条浪漫的蜿蜒大路展开，串联起社区内一系列的小型邻里；它们有些封闭，有些开放，使得漫步的人们可以自由地走街串巷。小镇中心以及位于小镇外围、运河沿岸的小尺度办公区均可提供就业岗位，这样就可能在步行可达的区域内，同时满足工作和生活的需求（图12，图13）。

在尺度上，泰晤士小镇很像英国的乡村小镇；但与中国其他风格主义的项目不同，它不是简单地对普通房屋进行立面装饰。不同寻常的城镇结构似乎给它的市场营销带来很大困难，直到小镇建成开放若干年后，这里的商业和办公建筑仍有大量空置。英国阿特金斯顾问公司是泰晤士小镇的设计方，他们根据既定的理念框架，提出一个颇为有趣的设计方案。将设计效果图和最终建成的效果进行对

Unlike most new development areas, Thames Town seems filled with people, walking the streets, eating in restaurants, and shopping in the few shops that seem to have survived beyond their initial opening. Most of all, it is the desired setting for wedding photography. On a sunny day, dozens of couples may be seen posing in their rented costumes, in the streets, on the lawn of the church, along the lake and canals, and in any picturesque corner of Thames Town. The stage set quality of the town has been recognized by the several wedding photography businesses that have displaced boutiques on the main streets (Fig. 10, Fig. 11).

What makes the town so attractive? Its scale is one factor-buildings that are human scaled in contrast to so much new development that towers above those on the ground. It's also the mixture of uses, having public buildings, churches, houses, and shops all very close to each other. It's the texture of small green spaces, squares and passageways, or as well as private green areas. The plan of the community is organized around a romantic winding road that connects a series of smaller neighborhoods within the community. Some of these are gated, although others are open and allow strollers to make their way from block to block. There is employment in the town center and in a low-scaled office district surrounding a canal at the edge of Thames Town. Thus it is possible to live and work within a pedestrian walking district (Fig. 12, Fig. 13).

In scale, Thames Town is very much like an English country town. Unlike other mannerist projects in China, it is not simply a series of facades pasted on ordinary buildings. The unusual structures seem to have to created marketing difficulties, and several years after its opening, the commercial and office space seems largely vacant. Atkins, a British consulting firm, designed the development. Within the framework they chose, they have done quite an interesting project. Comparing their renderings with the completed project, they have produced a community that is true to its intentions.

图8 纽约森林山村
（资料来源：Gray Hack）

Fig.8 Forest Hills Village, New York
(Source: Gray Hack)

图9 泰晤士小镇市民广场
（资料来源：Gray Hack）

Fig.9 Civic Square, Thames Town
(Source: Gray Hack)

图10 泰晤士小镇
（资料来源：Gray Hack）

Fig.10 Thames Town
(Source: Gray Hack)

比可以看出，阿特金斯创造出了一个忠于原创的社区。

遍布各处的雕塑也在不断彰显着泰晤士小镇的灵感之源，其中既有温斯顿·丘吉尔的铜像，也有查理·波特及其同僚的塑像。年轻一代的新都市人则聚居在小镇的另一区域。这里的保安都穿着英式制服，社区里有很多英国主题的商铺、小酒吧和其他销售点。

显然，这里也面临着如何适应中国生活方式的诸多问题。小镇中心的大多数住宅没有私人室外空间，居民们只得借用公共场合晾晒衣物和进行其他家庭活动。许多商店的门面也缺少典型中国店铺的繁华，难以在春节和其他传统节日时进行装饰。尽管因为目前居民数量有限，无法支撑大面积的商业，但与传统习俗不相适应可能也是造成店铺租赁困难的原因。另一方面，这里又有许多看似属于高收入家庭的大型门禁社区。

对此，我们能得出什么结论呢？未来，泰晤士小镇是否能够回应传统的期待？抑或像纽约皇后区的森林山小镇一样，因其个性鲜明和与众不同而成为连年升值的地方？又或者一直作为满足大众好奇心的参观之地和婚纱摄影留念胜地？显然，现在做出判断还为时过早；中国社会还处于快速变化中，也许一部分年轻的专业人士会在这里找到归宿。

3.4 安亭新镇

来到安亭新镇，会立刻感到新奇，似乎到了一个德国的新建住区，到处都是完全开敞的街道，却在边界处增设了检查站。每个进入新镇的人员或车辆都会被仔细盘查，显示了开放城镇理念与管控出入意愿之间的对立。也许，这项措施是暂时的，因为目前新镇的许多房屋还处于空置状态；当

The references to its origin are reinforced by the prevalence of statues throughout the community. Winston Churchill may be found there in bronze, as well as Charlie Potter and his colleagues. Ensembles of young urbanites inhabit another area. Security guards are dressed in British costume, and there is an abundance of British themed shops, pubs and other outlets.

Clearly, there are some issues of fit with Chinese life styles. Most of the town center houses have no private outdoor spaces, forcing the use of public spaces for clothes drying and other family activities. Many of the shop fronts lack the kind of exuberance of typical Chinese shops. The area isn't easily adorned for Chinese New Year and other celebrations. The lack of fit with conventions may account for some of the difficulty in leasing spaces, although the small resident population doesn't provide the base for many shops. On the other hand, there are many large gated estates that seem to be occupied by upper income households

So what can you say about this? Will Thames Town be adapted to meet conventional expectations, or is it going to be like Forest Hills, a place seems to be more valuable each year because it has a distinct character and differs from other alternatives? Or will it remain a curiosity for visitors and a stage set for wedding photographs? Clearly it is too soon to judge. Chinese society is changing quickly, and a fraction of the young professional class may find it to their liking, indeed.

3.4 Anting New Town

When you arrive in Anting New Town, there's an immediate curiosity designed as a new residential neighborhood that might be in Germany with completely open streets, a checkpoint has been added at the edge of the town. Every vehicle and person entering the time is scrutinized, exemplifying the tension between the idea of an open town and the desired control over who enters. This may be temporary because not many of the units are occupied, and perhaps it will be taken down when more of the units are occupied and the streets are filled with more activity. Or perhaps it will be the precursor for a more heavily gated development, like most Chinese residential areas (Fig. 14).

图 11　泰晤士小镇
（资料来源：Gray Hack）

Fig.11　Thames Town
(Source: Gray Hack)

图 12　泰晤士小镇大教堂
（资料来源：Gray Hack）

Fig.12　Cathedral, Thames Town
(Source: Gray Hack)

更多房屋投入使用，街道上有更多活动发生时，检查站就会被取消。又或者，目前的状态只是日后更加严格防卫的住区开发的前奏，就像大多数中国住区一样（图14）。

和泰晤士小镇一样，安亭新镇是更大规模的嘉定新城的重要开发区域。嘉定新城中心有便捷的公共交通，地铁车站周边的大型购物综合体也已经建成。嘉定旧城同样拥有热闹的商业街到、新建的公共广场和文化设施，以及夜生活丰富的古运河商业区。在此前提下，安亭新镇面向的则是中产阶级和专业人士家庭，他们的数量正在不断增加，并且正在寻找适宜的居住环境。虽然安亭新镇并非针对上述市场的唯一开发区域，还有许多其他地方都在推销拥有大片开敞空间的奢华封闭社区，但是安亭新镇却是其中唯一一个依照欧洲家庭邻里模式设计的开发项目（图15）。

大众汽车与通用别克都在嘉定的历史中心设有各自的行政总部。毗邻安定新镇则建起了一座面向汽车工业的大型博览中心，一座令人印象深刻的汽车博物馆，一个汽车设计中心，面向汽车产业的办公设施，以及囊括了所有在华汽车销售品牌的汽车展销中心，以方便顾客货比三家。在一站地铁之外，还建起了庞大的F1赛车场地和娱乐中心，延续着汽车城的主题（图16，图17）。

在安亭新镇，典型街区由沿街布置的4～6层住宅组成，底层用于商业和办公，背后是停车场和半私密的开放空间。有些内部庭院被封闭起来，仅供住户使用，有些停车场则被安排在地下。目前，许多商铺还处于空置状态，最活跃的地方似乎是占据了核心位置的老年活动中心。住宅也同样只有部分入住，因为缺乏私人室外空间，为数不多的住户也不得不占用共用的室外空间晾晒衣物、清洗车辆，以及完成其他必须在室外进行的活动。很难判断高空

Like Thames Town, Anting New Town is one signature development in the much larger city of Jiading. Downtown Jiading is well served by transit, and a large new shopping complex has recently been completed adjacent to the Metro stop. Older portions of the city have active commercial streets, new public squares, cultural facilities and a historic canal market district filled with nightlife. Anting New Town appears aimed at the increased number of middle class and professional families that are looking for a hospitable residential environment. It is not the only large new development aimed at this market; other areas advertise luxury, gated communities, with large areas of open space behind the walls. But Anting New Town is the only new development designed in the manner of a European family neighborhood (Fig. 15).

Both Volkswagen and GM Buick have their administrative headquarters in the historic center of Jiading. A large automotive-oriented exposition center has been created adjacent to Anting New Town, along with an impressive automobile museum. Also nearby is an automotive design complex, and offices for businesses in the automotive industry. The area also includes impressive sales centers for every automobile brand available in China, making comparison-shopping easy for consumers. Further afield, one Metro stop away, is a huge Formula 1 racetrack and entertainment complex, extending the theme of automobile city (Fig. 16, Fig. 17).

The typical block in Anting new town consists of 4-6 story housing lining the street, with commercial and office uses on the base. Parking is located behind, along with semi-private open space. Some of the internal courtyards have been gated to serve just the residents, and some of the parking has been created in garages below ground. Currently, many of the shops are vacant – the most active shop front seems to be the elderly activity center occupying a key space – and only a fraction of the housing seems to be occupied. With no private outdoor space the few current residents have taken to using the shared outdoor space for drying laundry, washing cars and other activities that need occur outside their homes. It is difficult to determine whether the high vacancy rate is a marketing issue (too expensive, not the kind of accommodation being sought, poor marketing campaign?) or a problem with

图 13　泰晤士小镇科研办公综合楼
（资料来源：Gray Hack）

Fig.13　Research and Office Complex, Thames Town
(Source: Gray Hack)

图 14　安亭新镇入口
（资料来源：Gray Hack）

Fig.14　Entrance to Anting New Town
(Source: Gray Hack)

图 15　嘉定新城中心
（资料来源：Gray Hack）

Fig.15　Jiading City Center
(Source: Gray Hack)

图 16 安亭汽车博物馆
（资料来源：Gray Hack）

Fig.16 Anting Automobile Museum
(Source: Gray Hack)

图 17 安亭汽车创意中心
（资料来源：Gray Hack）

Fig.17 Anting Automotive Innovation Center
(Source: Gray Hack)

图 18 安亭新镇典型街区
（资料来源：Gray Hack）

Fig.18 Typical Block, Anting New Town
(Source: Gray Hack)

置率是营销问题（例如价格过高，产品定位不准确，或者市场宣传不够？）还是设计问题，答案需要时间来揭晓（图18～图20）。

安亭新镇的最奇特之处是被命名为魏玛广场的镇中心。它由环绕巨大公共广场的市政大楼、公共市场、教堂、文化中心、办公楼和住宅楼组成，意图成为安亭新镇及其周边地区新的市民中心。然而，它的尺度与其功能严重偏离，尽管周围的房屋早在一两年前就已建成，但至今仍然空置；人行道上杂草丛生，显现出早期的衰退迹象（图21）。

安亭新镇的规划建设显然受到以沃尔夫斯堡为代表的德国北部工业城市的影响。虽然前身是一座历史村落，但沃尔夫斯堡在德国是相对年轻的城市，是生产"人民的汽车"高涨时的产物。多年来，其规模不断扩大，新建了几座博物馆，其中包括阿尔瓦·阿尔托的作品，但其基本空间结构仍以小规模的住宅街区为主，沿街布置有商店，围绕古老的沃尔夫斯堡广场形成了充满活力的市镇中心。在城墙被拆除之后，取而代之的是一条绿带，这一细节被原封不动地复制到安亭新镇。许多北欧城市都保留有城墙，但是沃尔夫斯堡除外，因为它大约只有百年的历史，相对年轻。所以沃尔夫斯堡拆除了城墙，并将原址建为公园，以限定城市的边界，由此形成大片林地公园环绕城市的景象，无疑成为施佩尔及其合伙人事务所规划设计安亭新镇环城绿色空间的重要参考。

和泰晤士小镇一样，遍布各处的雕塑强化了安亭新镇与其欧洲原型城市之间的联系。哥德的雕像就位于公共场合的显著位置。然而，这里的公共空间缺乏中国城市常有的休息场所和社交空间；特别在白天，街道和公共空间

the design. The answer will take time to determine (Fig. 18–Fig. 20).

The most curious feature of Anting New Town is the new town center, labeled Weimar Square. Built around a massive public square, that has a civic building, public market, cathedral, cultural center, offices and housing around it, Weimar Square is intended to be the new civic square of at least this portion of Anting. But it is massively out of scale with its purposes, and none of the buildings seem to be empty despite their completion a year or two ago. Grass and weeds have forced their way thorough the pavement, creating the impression of the early stages of decay (Fig. 21).

Clearly North German industrial towns such as Wolfsburg have influenced Anting New Town. Wolfsburg is a relatively young town, the product of the growth of production of the "people's car", but built on the base of a historic village. Over the years, extensions have been built, museums added (including one designed by Alvar Aalto), but its basic structure is small blocks of housing with shops lining streets. A greenbelt replaced the town wall when it was removed, a detail that was faithfully reproduced in Anting New Town. A lively and active town center surrounds the historic square of Wolfsburg. In many north European cities- Wolfsburg is not one of them because it's a relatively young city, maybe a hundred years old- were built walls around them. And they took the walls down and they preserve the area of the walls as a park, which becomes an outer perimeter of the town. Large woodland parks surround the town, which Albert Speer and Partners were undoubtedly mindful of as they planned the green surroundings of Anting New Town.

As in Thames Town, sculptures reinforce the connections between the new town and its European counterparts. In this case, Goethe has an honored location in the public realm. But public spaces in Anting New Town lack sitting areas and places for socialization usually associated with Chinese cities. Many of the occupants of the streets and public spaces are elderly, especially during the day, and they are forced to bring their own folding chairs down to the street, or use seating temporarily added to tree supports. The tradition of grandparents looking after the young children during the day, which is common in two

里多为老人，他们不得不自带折叠座椅，或者干脆坐在靠树干支撑的临时座椅上。在中国的双职工家庭中，爷爷奶奶在白天照顾孩子的传统非常普遍，这应该在公共空间设计中加以体现。安亭新镇的规划师者们显然忽视了这一点（图22）。

某些设计不当之处可以随时间的推移得到修正，使社区更好地服务于其居民，但魏玛广场的前途会是怎样？它能否达到德国北部城市广场质量水平呢？魏玛广场被设计成一个现代广场，就像在今天的沃尔夫斯堡修建一个新的市民广场一样。它其实又像美国的林肯中心，具有某种文化中心的职能。但是，与安亭新镇的规模相比，魏玛广场的尺度大而无当，也许应该考虑引入其他活动来占用部分空间。

3.5 浦江新镇

浦江新镇是关于上海郊区新镇的第三个案例，也是一个把城市和规划的思考方式跨文化地移植到另一种文化上的极端案例。浦江新镇位于2010世博会场址以南，其规划是国际竞赛的产物，由格里高蒂国际联合公司主导完成，具有当地城市开发的普遍特点，几条带状开敞空间与河流相连，将城市开发分隔成若干高密度组团。然而，由于受到居住在河边的当地村民的抗议，规划方案未能完全实施。尽管如此，在新建地铁8号线周边的农田上，已经出现相当的建设规模。和前文讨论过的其他案例一样，在地铁车站和浦江新镇的住区之间缺乏公交联系，并且更加糟糕的是，浦江新镇的大部分地区被宽阔的高速公路干线和带状绿地从地铁车站隔开。

working parent households, needs to be reflected in the design of public spaces. Apparently the planners of Anting New Town were unaware of it (Fig. 22).

Some of the misfits can be altered over time as the community is adapted to serve its occupants better, but what will be the fate of Weimar Square? Will this ever have the quality of the squares in North German cities? It has been designed as a modern square, as it might have been built if you were building a new civic square in Wolfsburg today. It actually harkens back to Lincoln Center in the United States, as a kind of cultural center. But it seems completely out of scale with the modest size of Anting New Town. Perhaps other activities can be found to occupy some of the space.

3.5 Pujiang New Town

The third example of a new town on Shanghai's periphery is actually an extreme example of transplanting a way of thinking about cities and planning to a foreign culture. Pujiang New Town was the subject of a competition, which selected Gregotti Associates as the lead designer. Located to the south of the Expo 2010 site, it has some of the characteristics of that development – dense urban clusters separated by ribbons of open space that connect to the river. However, resistance from local villagers living in communities along the river meant that the full plan has been impossible to implement. Nonetheless a substantial development is now taking form on former agricultural lands adjacent to the new Metro line #8. As with the other cases discussed, there is poor transit service between the Metro station and the residential areas. Making matters worse, the main sections of Pujiang New Town are separated from the transit by a broad swath of arterial highway and linear green spaces.

图19 安亭新镇
（资料来源：Gray Hack）

Fig.19 Anting New Town
(Source: Gray Hack)

图20 安亭新镇公共拱廊内晾晒的衣服
（资料来源：Gray Hack）

Fig.20 Clothes Drying in Public Arcade, Anting New Town (Source: Gray Hack)

图21 安亭新镇的魏玛广场
（资料来源：Gray Hack）

Fig.21 Weimar Square, Anting New Town
(Source: Gray Hack)

图22 安亭新镇的公共空间
（资料来源：Gray Hack）

Fig.22 Public Space, Anting New Town
(Source: Gray Hack)

图23 SWA 和斯卡凯迪的浦江新镇规划方案
（资料来源：SWA 事务所，Scaccheti 事务所）

Fig.23 SWA and Scacchetti Schemes, Pujiang New Town (Source: SWA Associates, Scaccheti Associati)

图24 格里高蒂的浦江新镇规划方案
（资料来源：Gregotti 国际事务所）

Fig.24 Gregotti Plan, Pujiang New Town
(Source: Gregotti Associati International)

在浦江新镇规划最终定案之前，参加竞赛的美国团队SWA曾提出基于汽车交通的蔓延开发方案，堪称加州橘子郡在上海。卢卡·斯卡凯迪领衔的意大利设计团队提出一个真正意义上的拼贴城市规划方案，把数个欧洲城市的片段缝合在一起。最终的实施方案来自意大利的格里高蒂设计团队，拥有更为有趣的特质。他们提出，一期开发原封不动地保留现有村落，只在地铁线路与东部就业中心和未来的沿江绿带之间建造一系列带状社区，把拆除村落作为整个开发的最后环节。然而他们没有想到，如此必然带来村落升值，这也几乎决定了它们可以永久地留在原处（图23，图24）。

设计团队就新镇街区的适宜尺寸进行了大量分析，最终决定把300米×300米的模块作为基本单元。该单元又可以被继续细分为四个150米×150米的地块。这种尺度的街区可以确保沿街区周边布置不同尺寸的房屋，同时街区内部还保留充足的绿地空间。设计师们对街区内部的可能布局也进行了大量实验，但我认为这些研究都是徒劳，因为没人能真正了解生活在硬质边缘的场所与生活在软质边缘的场所有何区别，也无人知晓人们的生活习惯会有怎样的改变。不管怎样，设计师们从这些研究中得出结论，在150米×150米和300米×300米的模块内几乎有无限的布局可能；这样的形态学分析似乎成为意大利未来主义思潮的风尚。

在一期开发规划设计进程中，最先确定的决策之一便是将机动车全部停放在地下，这虽然代价昂贵，但重要的是可以将地面解放出来，进行更为灵活的规划。设计团队提出，在东西方向上，利用内部公共空间将各个街区串联起来，沿线布置店铺与机构，中间点缀以广场空地。这显

The plan for Pujiang New Town was not preordained. SWA, a US competitor, proposed a spread development relying mainly on auto travel – Orange County California in Shanghai. An Italian team headed by Luca Scacchetti submitted a plan that was literally a collage city, cobbling together fragments from a number of European cities. The plan selected was by Gregotti, Associates also from Italy and it had an interesting quality. They proposed leaving the villages intact in the first phase, while building a series of linear communities that connected the Metro line and employment centers to the east with the future greenbelt along the river. Demolishing the villages would be the last stage of the development. Little did they know that by making the villages even more valuable, they virtually assured that they would remain in place (Fig. 23, Fig. 24).

The design team did many analyses of the appropriate size of a development block, finally arriving at a notion that the basic module ought to be 300 m by 300 m, which could then be subdivided into four quadrants 150 m by 150 m. Blocks of this size provided enough room for internal green spaces with buildings of varying sizes surrounding them. They performed an endless number of exercises of the internal arrangements of blocks that are possible, studies that I consider to be misplaced energy. I say this because nobody really knows the difference between living in a space that has a hard surface in the center and a place that has a soft surface in its place; nobody knows how peoples living patterns would be altered. Nonetheless, it allowed the designers to conclude that almost infinite variety was possible within 150 m by 150 m module, and 300 m by 300 m blocks. Such morphological analyses seem to be de rigeur in Italian futurist thinking.

One the of the first decisions made in designing the first phase development was to put all of the automobile parking underground. This was a large and costly decision, but it allowed the ground surface to be planned with more flexibility. Gregotti Associates proposed connecting blocks in an east-west direction via an internal public spaces, lined by institutions and shops, and punctuated by squares along the way. The image is of an Italian town, where it is possible to walk the length of the town enjoying meeting people and window-

然是意大利城镇的场景，人们可以徒步横穿整个城镇，沿途见面寒暄、浏览橱窗，直到最后进到餐厅就餐。最初，浦江新镇的街区都依此模式建造，最终结果却与预期相去甚远，因为现实中每个住宅组团都希望通过建造围墙把自己与外界交通分隔开来。为了阻止路人闯入街区的内部绿地，人们在街区之间设置了围栏岗亭。这无疑是文化理解差异而导致的重大败笔（图25，图26）。

公共空间的割据还导致了其他一些后果，包括缺少足够的顾客光顾步行街沿线的商铺。在整个开发过程中，商店生意举步维艰，只有大型超市和健身俱乐部能吸引到足够的消费者。公共空间也大都空空荡荡，而且就像前文所述的其他新镇一样，浦江新镇的市场营销进展比预期要缓慢许多。

浦江新镇里最成功的开发似乎是被围墙环绕的低密度别墅区。这里提供了多种类型的住房，多数拥有室外私人空间。有些住房面向运河，把原本向公众开放的运河步道据为己有。这再次表明，这里不是威尼斯（图27，图28）。

除了早期建设的密度较低的居住区外，这里还建起了一系列文化教育机构，包括画廊、教育综合楼和中意广场（China-Italy Place）。这个意大利未来主义风格的广场有着过于浓重的仪式感，仿佛是为一些重大活动专门准备的，能否吸引足够的人群使其成为真正的核心，至今仍是一个问题。也许，随着未来的开发密度越来越更高，慕名前来定居的人越来越多，这里会变得更加充满生机（图29，图30）。

3.6 几点结论

前文列举的几个案例表明，风格主义城市面临的风

shopping before dining out in a restaurant. The initial prototype blocks have been constructed in this manner. However, this effort has come to naught, as each housing cluster sought to wall themselves off from outside traffic. Fences have been erected between blocks and guard posts have been established to prevent passersby from intruding on the green spaces of each block. This is a major failure of understanding cultural differences (Fig. 25, Fig. 26).

The balkanization of the public realm has had other consequences, including the lack of adequate patronage to keep shops along pedestrian walkways alive. Throughout the development, shops seem to be struggling, with only the largest supermarkets and health clubs apparently attracting adequate customers. Public spaces are also largely empty, and like the other new towns just discussed, marketing of Pujiang appears to be much slower than anticipated.

The most successful areas of Pujiang New Town appear to be the low-density villa areas, separated from the street by walls. Many variations of housing are offered, and most have outdoor private spaces. Some of the housing units front on canals, and some of the canal walks intended to be open to the public have been expropriated by the adjacent housing owners. Venice this is not (Fig. 27, Fig. 28).

Beyond the initial residential areas, which are very modest densities, a series of cultural and educational institutions have been constructed. Among them are an art gallery, education complex and a public space called China-Italy Place. It also suffers from excessive formality just noted, and resembles a kind of Italian futurist square awaiting some major events. Whether it will ever attract enough people to make it a central place is an open question. Future developments will be much higher densities, and perhaps as more residents are attracted to Pujiang New Town, the center will become more animated (Fig. 29, Fig. 30).

3.6 Some Conclusions

Mannerist communities run the danger of misreading the cultures they are designed to serve, as each of the examples demonstrates. But cultures are not static, particularly in a

图25 浦江新镇的内部街道
（资料来源：Gray Hack）

Fig.25 Internal Street, Pujiang New Town
(Source: Gray Hack)

图26 浦江新镇的内部广场
（资料来源：Gray Hack）

Fig.26 Internal Square, Pujiang New Town
(Source: Gray Hack)

险是误读作为其服务对象的文化。文化并非一成不变，尤其在中国这样飞速变化的社会，城市建设需要服务于几代人；下一代人是否拥有与当代人相同的愿望？什么可以持久，什么又是昙花一现？伴随社会的进步，简单重复当前的生活状态同样面临被淘汰的危险。

那么，城市形态的哪些方面会持久？哪些又会消逝呢？这是城市设计者必须面对的关键问题。从前述实例可以看出，海外舶来的原型在应对产权和社会排斥等问题时力不从心。中国城市历来就有通过建设围墙将他人隔离在住区之外的悠久传统。在过去六十多年里，不少机构采用这种方式将自己孤立起来；在更为久远的历史上，每个家族的聚落以及每个少数民族的聚居区也都建有围墙。在欧洲和北美社会，人们则坚持通过抵制封闭街道或门禁社区的公共政策，努力打破隔墙和消除障碍；然而，看看加州门禁社区的数量，或者巴西城市异常严格的门禁管理就能知道，这样的努力并非总能获得成功。因此，要看到把开放城市理念引入中国的风险，要么是市场营销的失败，要么是后期加建的各种门卫和围墙，就像在上述案例中看到的那样。

街道生活是第二个值得探讨的话题。低收入阶层大都拥有占用街道的文化特点，部分原因是作为个体，他们少有其他社交方式可选。对于较高收入阶层而言，由于拥有私人娱乐场所和设施的选择，他们更倾向于将街道视为通行的场所。许多新兴中国城市都将街道设计为交通专用，一些借鉴欧洲理念的新城还将街道视为各种管网的通道，哪怕街道两侧商铺林立。塑造供人流连忘返、见面寒暄的公共场所是中国城市由来已久的特质，然而本文涉及的所有案例在这一点上都相距甚远。随着时间的推移，

rapidly changing society like China. Communities are built for several generations. Will the next generation have the same aspirations as the current one? What will be durable, and what preferences will be transient? Simply repeating current life styles and preferences runs an equal risk of being obsolete as a society evolves.

What aspects of urban form are likely to be durable, and which will be transient? That is the central question urban designers must ask. From the examples just discussed, the issues of property and exclusion seem to be poorly handled in models adapted from abroad. Chinese cities have a long history of constructing walls, and excluding others form residential areas. Over the past six decades isolating institutions has done this, but over the longer history walls were created around individual family compounds and ethnic enclaves. European and North American societies have sought to break down walls and remove barriers to access – often through public policies against closing streets or gating communities. This has not always been successful – witness the number of gated communities constructed in California, or the extreme gating of Brazilian cities. But transporting the logic of opening up cities in China runs the risk of either marketing failure, or later additions of walls and gates, as we have seen in the examples above.

Street life is a second contentious issue. Low-income societies are almost universally street using cultures, partly because individuals have few other choices for socialization. Higher income societies that have the choice of private property and institutions for recreation view streets more as conduits for movement. Many new Chinese cities are being designed as if streets are exclusively for movement, and several of the new towns borrowing European ideas regard streets as largely utilitarian channels, even if they have shops lining them. Creating places to linger, meet others, and be in public would seem an enduring quality of Chinese cities, yet all of the examples described here fell far short of what is needed. Will Chinese patterns of street use become more European or American over time? My guess is that the venues sought will change, but there will be no diminution of desire for being in public.

图 27　浦江新镇的低密度住宅
（资料来源：Gray Hack）

Fig.27　Low Density Housing, Pujiang New Town
(Source: Gray Hack)

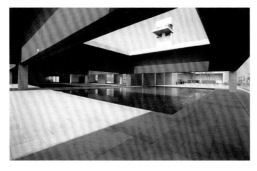

图 28　浦江新镇的运河开发
（资料来源：Gray Hack）

Fig.28　Canalside Development, Pujiang New Town
(Source: Gray Hack)

中国的街道使用模式是否会欧洲化或美国化？我的猜想是，人们对场所的要求也许会改变，但融入公共场合的渴望不会消逝。

设计能够因随时间不断变化的城市是应对社会更迭的有效方法，设想最初构思如何应变与满足当下需求同等重要。观察其他社会的城市形态当然可以有所收益，但是缺乏批判的原样照搬一定是得不偿失。

Designing cities that have the capacity to change over time can serve as a hedge against social change. Imagining how the initial pattern may be adapted is as important as accommodating current needs. Much can be learned from observing the urban forms of other societies, but uncritical borrowing of ideas runs the risk of missing the mark.

图29 浦江新镇中意广场
（资料来源：Gray Hack）

Fig.29 China-Italy Place, Pujiang New Town
(Source: Gray Hack)

图30 浦江新镇中意广场视野
（资料来源：Gray Hack）

Fig.30 View from China-Italy Place, Pujiang New Town
(Source: Gray Hack)

参考文献 / References

ANGELO S, GOVERNA F, BONINO M, 2018. Exploring Chinese New Towns. https://urbannext.net/exploring-chinese-new-towns/.

BIANCA B, 2013. Original Copies: Architectural mimicry in contemporary China. Honolulu, University of Hawaii Press.

JENNIFER J, 2013. The urban now: Theorising cities beyond the new. European Journal of Cultural Studies 16 (6): 659-677.

山地生态
HILLSIDE ECOLOGIES

盖里·海克
宾夕法尼亚大学设计学院荣退教授
清华大学建筑学院客座教授

Gary Hack
Emeritus Professor, School of Design, University of Pennsylvania
Guest Professor, School of Architecture, Tsinghua University

中文翻译：梁思思，刘健
Chinese translation by Sisi Liang and Jian Liu

近期，我受2012第一届重庆大学城市可持续发展国际学术会议暨第三届山地人居科学国际论坛邀请，做了关于山地可持续发展的学术报告。必须承认，我在平缓的加拿大西部地区长大，那里盛产小麦和其他作物，因为地势非常平坦，从不同方向都可以远眺地平线。所以对我而言，任何山地，不论尺度如何，都是特别的事物；我会时常想起山脉与丘陵，因为在任何成长于平原地带的人眼里，它们都是那么美好。

从另一个角度看，山地也十分重要。据统计，全球约有30%的人口居住于山地地区，其中包括很大一部分的中国人口，因为中国有很大面积的国土属于丘陵和山地；例如在香港地区，除了填海的区域外，大部分城市都建造在山坡上。然而，当下盛行的多数城市设计理念都假设土地是平整的，人们可以在上面随心所欲地布置道路，由此形成的空间形态常常取决于建筑形式而非自然地势。

在此，我想分享一些关于山地开发的思考，以及世界各地有意思的山地开发案例，希望它们能够激发更多的想法和理念，促进山地城市设计的发展。

4.1 山地生态

关于山地生态，意大利的波西塔诺，这座位于那不勒斯南部阿尔玛菲海岸线上的历史名镇，是可以参考的案例。人类在此定居已有约600年时间，山地开发从海岸一直延伸到山坡顶部，成为著名胜地。几百年来，渔民们始终将船只停靠在岸边，但城市目前的主要经济来源则是旅游业，各种车辆通过蜿蜒曲折的道路，从高而低抵达小镇各处（图1）。

Recently, I was asked to give a talk at a conference in Chongqing on sustainable development in hillside areas, i.e. The 1st Chongqing University International Symposium on Urban Sustainable Development and The 3rd International Symposium on Science of Human Settlements in Mountain Region. I have to confess that I grew up in a very flat part of the world, in western Canada where they grow wheat and other grains. It is so flat there you can see the horizon in all directions. So for me, a hill of any size is a special event. I think about mountains and hills quite a lot because they are a great pleasure for anyone who grows up in the flatlands.

Hillsides are also a very important subject form another standpoint: by some estimates about 30% of the world's population lives on hillsides, including a large fraction of China's population, because so much of the country is hillsides, hilly, and mountainous. Hong Kong is largely built on mountainsides, except for land reclaimed from the sea. And yet most of the urban design ideas promoted today assume that you have a flat plain and that you can plan roadways wherever you want them to go. The typologies are driven by built form rather than natural typography.

What I want is to share thoughts about hillside development and examples of what I consider to be interesting hillside development in various places in the world. Hopefully they will stimulate other thoughts and ideas to advance the urban design of hillsides.

4.1 Hillside Ecologies

As a way of thinking about hillside ecologies, consider Positano, Italy, a historic town on the Amalfi Coast, south of Naples. People have inhabited this hillside for about six hundred years. It is a remarkable place, with hillside development extending from the beach up to the top of the hills. Fishermen continue to land their boats at the beach as they have for hundreds of years, but the economy of the city today is largely based on tourism. The town is reached by vehicles at the upper elevations with switchback roadways providing access to much of the hillside (Fig. 1).

图1 意大利波斯塔诺
（资料来源：123rf.com）

Fig.1 Positano Italy
(Source: 123rf.com)

几个世纪以来，波西塔诺形成了独一无二的生态系统——一个薄层的生态系统，从遍布山坡的树根底部到房屋顶部，厚度大约只有20米。大部分树木是柠檬树与香橙树，因此小镇的产业之一就是生产被称为柠檬赛露（Lemoncello）的甜酒。在旅馆的遮阳露台上，游客们可以一边享受这种甜酒，一边欣赏美丽的景色。城市生活依据穿行于坡地的需求而布局，鱼市位于山脚，其他商业店铺则位于道路交叉口附近。

盛产柠檬的自然生态与依坡就势的生活方式亲密交织，树根起到固化土壤、阻止滑坡和吸收雨水的作用，而树木的景致、芬芳和荫凉又被用于旅游开发。这个结构独特的人工生态历经数年发展至成熟稳定，庇护着山地上独一无二的生活方式。

波西塔诺案例还证明，聚落形态必须遵守若干规则以适应山地形态。例如，所有房屋都要沿山坡等高线布局，道路亦然；街道在某个高度上延伸，然后通过坡道迂回至高处或低处；机动车道被限制在坡度约15%的坡地范围内，由此也决定了开设街道的范围。波西塔诺的绿化大多位于山坡的陡峭之处，植被环裹山坡以防止水土流失，并缓解高密度人工环境带来的压力。

山地生态是两种生态要素的交汇，即自然系统与人工造系统相互交织。山地人居环境要求建立一个既能反映自然需求，又可造就符合人类意愿与需要的生活模式的生态系统。最终，人们以不同于平原的生活方式栖息于山地之上，他们依托的自然环境也与平地原野截然不同（图2）。

当然仔细研究不难发现，人类栖息的所有生态系统都呈现出人力与自然之力的交汇。典型的城市设计策略会首先明确最为脆弱的自然地域，杜绝在其中或者在附近进行

Over the centuries, Positano has created a unique ecology. It is a thin ecology, perhaps only 20 m thick, from the roots of trees that hug the hillside to the tops of buildings. Most of the trees bear lemons and oranges, and one of the industries of the town is making a liqueur called Lemoncello. Hotels have shaded terraces where the liqueur is served to patrons enjoying the wonderful views. The life of the city is adapted to the necessity of traversing the hillside, with the fish markets at the base and other commercial outlets nearer the roadway access points.

The natural ecology of growing lemons and the way of life that has evolved on this hillside are intimately intertwined, because the roots of the trees stabilize the soil, stop landslides, and absorb the rainfall, and the trees have been exploited for their beauty, fragrance and shade to support tourism. So this is a unique constructed ecology that's taken many years to mature, but which is stable and accommodates a way of life that is uniquely suited to the hillside.

Positano also demonstrates that the morphology of a settlement must observe several rules to conform to hillsides. All buildings need to be oriented parallel to the hillside, and the roadways similarly so. Streets tend to hug a single elevation and then loop back to a lower or higher elevation, traversing the slope. Automobile roadways are limited to about a 15% slope, which becomes a determinant of where streets can be created. Green space in Positano tends to be on the steeper parts of the slope where the vegetation hugs the hillside stopping erosion and providing relief from the dense built environment.

Hillside ecologies, thus, are the intersection of two kinds of ecological components: natural systems intertwined with human systems. Inhabiting a hillside requires constructing an ecology which reflects the demands of nature, but also results in a pattern of life that is responsive to people's desires and necessities. Ultimately people will live differently on a hillside than on the flat, and the kind of natural world they will interact with will be different from wilderness (Fig. 2).

Of course, if you look closely, all ecologies inhabited by humans represent an intersection of human and natural forces. A typical urban design strategy is to identify the most fragile natural areas, and protect them by avoiding any development in or near them. But inevitably they change as water resources are

图2 山地生态概念模型
（资料来源：Gray Hack）

Fig.2 Conceptual Model of Hillside Ecologies
(Source: Gray Hack)

开发，以便对其进行保护。但是随着水资源的改变、动物群落被圈养、建筑产生微气候环境以及新的植物种类被引进，这些地区会不可避免地发生变化。因此需要建立一个人类—自然复合生态系统，而不是使二者相互隔离。这一点对于山地地区尤为重要，因为在那里，即使最微小的人类干预，例如修路、伐林，都会牵一发而动全身，带来不可逆转的后果。

山地的生态机制因高度、朝向与气候而异。在山顶，因为过于寒冷或陡峭，或者常年积雪覆盖，常常少有天然植被；当山顶的冰川融化，便为低处的植被生长创造了条件。沿山坡而下，在雨水充沛的向阳山坡，可以发现更多茂盛的植被；在河流两岸，也有更加高大的树木吸收土壤营养与山谷积水。因此，山地并非单一的生态系统，而是一系列持续变化的机制（图3）。

有时，微小的地理差异也可能造成生态机制的极大不同。例如，在一片贫瘠的土地上，一条小小的山谷就可借由山地的阴翳和山上偶然流下的水流养育树木，而光照更加充足的南向坡地也可以养育与相邻的朝东或者西南朝向地区完全不同的植物。在进行景观分区时，对诸如此类的微妙之处都要有所考虑。种植葡萄酿酒的人们深谙此道，因为在不同坡地上种植的葡萄种类会有很大不同。因此，在规划山地聚落时，这些潜在的自然要素都是需要考虑的问题。

当人类削山造地或者当大火在山间蔓延时，人与自然的演替过程开始出现；人类侵蚀破坏或改变山体，自然生长达到足够规模使其趋于稳定，二者之间最终形成竞争关系。人类的干预越强，例如在山地上建造房屋或者开垦耕种，人类需要应对的危害也越多，这其中又有许多与水的

altered, fauna are hemmed in, buildings create microclimates, and new plant species are imported. So the problem is one of constructing a human-natural ecology, not separating the two. This is especially true in hillside areas, where even small human interventions – building a road, clearing an area of forest – can set in motion a series of changes that are irreversible.

Mountainside ecological regimes vary with elevation, orientation, and climate. At the top of a mountainside, there is often no natural vegetation because it is too cold or too steep, or covered much of the year with snow. As the glaciers melt, they provide the basis for vegetation on the lower slope. Moving down the slope, as rainfall is coaxed out of the sky by mountainsides, you find more lush kinds of vegetation, and along rivers, large trees that draw their nourishment from the soils and water deposited in the valley. Thus, a hillside is not one ecology but a series of ever changing regimes (Fig. 3).

Sometimes small geographic differences can make a great deal of difference in the ecological regime. A small valley will nurture trees in an otherwise barren landscape, because of the shading of the hills and occasional water that flows down the hillside. A south sloping slope with more sunshine will support growth that is different from adjacent areas oriented just east or west of south. In categorizing landscape zones all of these subtleties need to be accounted for. People who grow grapes for wine really understand such subtleties, since the kind of vines that you can grow on one slope are very different from what can grow on others. These potentialities are what nature brings to the equation, when you are planning settlements on hillsides.

When man intervenes and clears fields or when fires burn the mountainside, a succession process is set in motion. And there is a race in time between when erosion destroys or alters the hillside and the nature grows up to a size that is capable of stabilizing it. With larger interventions, such as adding housing or agriculture to a hillside, there are many more hazards that one has to deal with — many of which are rooted in handling water. Water is both the life-blood of vegetation on hillsides, but also a life threatening challenge. Mountains are constantly being reshaped by water. It often undermines surface soils causing slippages, which as one looks closely at

图3　山地生态
（资料来源：Wikimedia）

Fig.3　Mountainside Ecologies
(Source: Wikimedia)

治理相关。水既是山地植被的生命之源，也会威胁人类生命，因为山地的形态不断地因水而变。仔细观察山区地貌可以发现，水经常破坏山地表面土层造成滑动，而严重的滑坡有时会使修建在错误地点的聚落荡然无存，诸如干涸多年的河谷以及水位超过堤坝高度的地区（图4）。

4.2 流

在人类在自然框架内作为固然重要，但更为重要的是，要认识到人类在栖息于山地时的一举一动实际上都在改变着生态环境。这便迫使设计师必须思考，如何为未来创造一个稳定的山地生态。

对此，有效的思考方法是通过绘图分析各种流，那些人们必须面对的逆坡而上的流和顺坡而下的流。有哪些流是顺坡而下的呢？答案是雨水、雪融水（如果有雪的话）、人流、产自山上的物品、污水、垃圾及其他。同样，逆坡而上的流包括饮用水（如果在高处没有足够水源的话）、车辆燃料、人流、住房及其供热和制冷、食物、交通工具、应急设施、建造材料等。所以，要同时考虑两个方向的流。山地城市设计的挑战就在于如何容纳和协调上述各种流、需要使用的技术类型、以及山地建成环境形态将它们融合在一起的方式。

由于山地的复杂性，历史上在绝大多数情况下，人们在坡地上谨慎行为，将聚落分散布局，或者布置于山顶，以分散排水、人流、车辆、道路和房屋所产生的影响。在希腊克里特岛卡兰村的古代聚落，各类建设在坡地上分散布局，成为这一方法的典型范例。同样，在中国的西藏和其他地区，首选的策略包括在坡地上分散布局，使构筑物

hillside topography are easy to spot. Sometimes, landslides are so dramatic that they sweep away settlements that have been built in the wrong place – in river valleys, which have been dry for many years, or areas where water overtops dams or dikes that have been created (Fig. 4).

4.2 Flows

Working within the natural framework is important, but more important is understanding that everything one does in inhabiting in hillside actually changes the ecology. This forces the designer to think about how to create a stable future hillside ecology.

The useful way to think about it is mapping the flows – flows of things up the hill and things down the hill that one has to deal with. What's going down the hillside? Rain water, snow melt (if you have snow,) people, products that are produced on the hillside, sewage, garbage, all of the other things finding their way down the hillside. Similarly, going up the hillside: drinking water, if there isn't enough in the upper elevations, fuel for vehicles, people, their houses and heating or cooling, food, vehicles, emergency equipment, building materials, and so on. So think of the flows in both directions. The challenge of doing an urban design for a hillside is how to accommodate and coordinate those flows, the type of technologies employed, and the way the morphology of the built hillside incorporates them.

Because of the complexity of the hillside, in most cases historically people have tried to tread lightly on hillsides, spreading settlement out or locating it on the top, so as to spread the impact of new drainage, people, vehicles, roadways and buildings. The ancient settlement of Karanou Crete exemplifies this approach, spreading structures across the hillside. Similarly, in China, in Tibet and other places, the preferred strategy has included spreading up the hillside, keeping structures low, constructing ways that natural drainage systems can flow through communities, all designed to minimize impacts (Fig. 5).

图4 中国舟曲山体滑坡
（资料来源：boston.com）

Fig.4 Landslide, Zhouqu, China
(Source: boston.com)

图5 克里特岛卡兰村
（资料来源：Greece.com）

Fig.5 Karanou, Crete
(Source: Greece.com)

图6 香港半山
（资料来源：Gray Hack）

Fig.6 Hong Kong Mid Level Development
(Source: Gray Hack)

图7 香港半山扶梯
（资料来源：Gray Hack）

Fig.7 Hong Kong Escalators to Mid Level
(Source: Gray Hack)

保持较低的高度，道路建设易于自然排水系统贯穿所有社区，凡此都是为了尽可能减小对环境的冲击（图5）。

这些策略在低密度条件下易于实现，但在中国的香港以及其他许多城市，例如重庆，则难以奢侈地采取谨慎的行为策略。由于有过多的人口需要安置，这些城市不得不面对如何在山地上有选择性地创造空间进行高密度开发的挑战。一种策略就是在山地构成中发掘自然形成的台地作为定居之所，进而通过技术手段将它们联系在一起，便于台地之间人与物的流动。在香港，高密度开发与高层建筑是一种非常合理选择，因为其山地的黏质土壤很容易受到滑坡和洪水的影响，因此建筑物的基础必须深入土壤直到稳固的承重地层；又因为深层地基造价昂贵，唯一符合逻辑的方法就是建造小基底面积的高层结构。为了连接中间和低层的台地，香港安装了一系列公共自动扶梯，它们迅速成为这个城市的新一代主要立体街道。这是面对特殊自然条件的一种不同应对方式，与在山地上采取谨慎行为策略不分伯仲。至此，人类便掌握了更具厚度的生态系统（图6，图7）。

在山地条件下历经长期发展的城市通常都会发明新的方式，实现人与物在山上和山下之间的运输。例如，在意大利热那亚，大部分的城市商业活动都在历史悠久的山脚滨水地区，因为这里是各种货物抵达的口岸，是鱼类、来自非洲的蔬菜和其他各种海上贸易集中交易的场所；而人们则选择在港口周边的坡地高处居住，因为那里更加凉爽，并且远离拥挤的中心，山上的习习凉风与陡峭坡地形成的丰富自然景观造就了令人向往的居住环境。

针对山上与山下之间的交通联系，热那亚的独特解决方案是一个面向所有人的公共升降系统，作为垂直的基础

These strategies can work in low-density situations. But in cases like in Hong Kong, or many other Chinese cities such as Chongqing you don't have the luxury of treading lightly on the hillsides. Many people need to be accommodated and the challenge is to create places for high-density development selectively on hillsides. One strategy is to seek out the natural benches in the hillside formations, settling these and connect them through technologies that smooth the flow of people and materials between the benches. In Hong Kong, high-density development and tall buildings make a great deal of sense because of the preponderance of clay soils on the hillsides; they are quite susceptible to landslides and flooding. So the foundations must go very deep into the soil to get to solid bearing material, and because of the expense of deep foundations, it is only logical to create tall structures with small footprints. Connecting the lower and mid-level benches, Hong Kong installed a series of public escalators that have quickly become the new vertical main streets of the city. This is a different kind of response to the very particular natural conditions they are facing, and it's not a better or worse response than sitting lightly on the hillside. A thick ecology can be mastered (Fig. 6, Fig. 7).

Cities that have evolved over long periods in hillside conditions have often invented new ways of transporting people and materials up and down the hillside. In Genoa, Italy most of the commercial life of the city sits at lower elevations near the historic waterfront, because that's where the port of arrival was, the fish were landed, vegetables came from Africa, and other kinds of maritime trades were practiced. However, people chose to live in the upper parts of the hillsides that surrounded the port, because it is cooler and away from the congested center. Breezes on the hillside, and the matrix of landscape that were necessitated by the steep slopes made for desirable residential districts.

Genoa's unique solution to movement up and down the slope is a system of public elevators, open to all. This constitutes vertical infrastructure, very much a part of public circulation system. There are several elevators at each of the major benches in the community. Sometimes, individual buildings make the transition privately as well, with entrances on both the

设施，成为公共交通系统的组成部分。城市中的各个主要台地都分别设置了多部扶梯；有时单体建筑也会通过设置在不同高度的入口，实现私密性的交通转换。热那亚独特的山地生态使其像平原城市一样便利（图8）。

适于耕种的土壤往往位于平原地区，也常常易于城市建设，只是一旦失去便不会再有。正是因为认识到这一点，许多城市都将视线转向山地，将其视为安置住房或其他形式开发的另外之选。例如在加拿大温哥华，弗雷泽河三角洲土地肥沃，长期种植水果和蔬菜，从1980年代开始严格限制城市开发，结果导致市区建设密度不断攀升，其中也包括了在周边山区的蔓延发展。西温哥华的山地开发展示了当城市建设顺坡而上时，城市发展如何不得不改变其形态。在坡度低于15%的山脚，城市主要采用了通常的网格道路；沿坡向上，道路形态逐渐转变为折返迂回的系统。同样，在地势低缓的山脚，流水会被引入管道；而在坡地高处，则需要将其露出来并在地表种植植被，以防造成土壤侵蚀，于是沿山涧形成的绿带就成为山地生态的组成部分。但是当坡度上升到25%～30%时，这些方法已无法适用，开发建设会变得异常昂贵并充满风险（图9）。

在加利佛尼亚的洛杉矶盆地，大部分平原地带已被占用，人们不得不通过采取特殊手段来创造可供开发的场地。对于山地，则意味着需要构建一个生态保护系统，防止土壤侵蚀与山体滑坡。对此，贝尔艾尔高地开发可以作为案例，说明为了稳定山体以便在山顶台地进行开发而采取的特殊手段，其中包括沿山涧密集植被并堆砌岩石以减缓其流速，在山坡上设置集水沟渠，以及在山脚开辟大面积蓄洪区。在洛杉矶，所有山地开发都必须建立这样的生态系统，防止城市开发因山体滑坡而遭到破坏。这就引发

upper and lower elevations. Genoa's unique hillside ecology makes it every bit as convenient as a flat land city (Fig. 8).

Good agricultural soils, usually on flat lands, are generally favored for urban development, but once lost, prime farmlands cannot be replaced. Recognizing this, many cities are looking to hillsides as an alternative for housing and other forms of development. In Vancouver, Canada, beginning around early 1980s, development was tightly restricted on the rich farmland used for growing fruits and vegetables in the Frazer River Delta. The result has been increased densities within the urban limits, including extensive development of hillsides of surrounding mountains. The hillsides of West Vancouver illustrate how the form of urban development has to change as it climbs mountainsides. At the base of the mountain, where slopes are less than 15%, a normal city grid predominates. Moving up the slope, the form of roadways shifts to a switchback system. Streams, which were traditionally captured in pipes on lower slopes, need to be exposed and vegetated to protect from erosion. Hence green ribbons lining the ravines become part of the hillside ecology. When slopes increase to 25%-30% they become too steep for even these devices, and development becomes prohibitively expensive and hazardous (Fig.9).

In California, where most of the flatlands in the Los Angeles basin have been occupied, extraordinary measures need to be taken to create development sites. Hillside ecologies need to be constructed to provide protection against erosion and landslides. The Bel Air Heights development illustrates the extraordinary measures needed to stabilize hillsides to allow the bench near the top of the hills to be exploited. These include densely planted ravines lined with rock to slow the flow, catchment trenches on the hillsides, and a large detention area at the base of the hills. In LA, all series of hillside ecologies have had to be constructed to protect them from being lost to landslides. There is a legitimate question of whether areas that require this extraordinary effort to be made safe should be developed, but if they are, they require a constructed ecology that is sustainable. Water is the great enemy, but also the great benefit that hillsides enjoy (Fig. 10).

图8 热那亚公共扶梯与垂直楼房
（资料来源：Franco Albini）

Fig.8 Genoa Public Elevators and Vertical Buildings
(Source: Franco Albini)

图9 西温哥华开发模式
（资料来源：Google Earth）

Fig.9 West Vancouver Development Pattern
(Source: Google Earth)

图10 洛杉矶山地开发（资料来源：洛杉矶城市规划局）

Fig.10　Los Angeles Hillside Development
(Source: Los Angeles Department of City Planning)

图11　旧金山伦巴第街（资料来源：Wikimedia）

Fig.11　Lombard Street, San Francisco
(Source: Wikimedia)

图12　旧金山电缆车（资料来源：Wikimedia）

Fig.12　San Francisco Cable Car
(Source: Wikimedia)

了一个问题，既然一个地方的开发需要采取如此特殊的手段才能保证安全，那么这个地方是否还应该开发，如果是，它们就需要构建起这样可持续的生态系统。在这样的生态系统中，水既是对山地的威胁，也可以为山地带来巨大收益（图10）。

旧金山等城市却是在人们提出重建生态概念之前即已建成。人们热爱旧金山，就因为城中陡峭的坡地。在这里，当开车爬上一个陡坡时，通常很难看到道路交叉口出现；因为按照当地规定，道路坡度不能超过20%，否则必须采用折返迂回的"之"字形方式，就像被称为"美国最蜿蜒街道"的伦巴第街。为了使车辆可以到达坡顶，许多山头都被削平；同时，在陡峭的街道上停车也很危险，所以停车位的方向都改为与道路垂直。坡度超过10%的人行道通常采用阶梯方式。当然，当地应对坡地条件的最伟大发明还是电缆车，利用埋在地下管道中的电缆将车厢拉上山坡，现已成为这座城市的标志（图11，图12）。

在世界其他地方，人们也发明并利用多种不同技术手段，将人员、物资和服务运上山顶，例如索道缆车、地上升降机等。在纽约，进入地处东河中游的罗斯福岛的主要方式是吊索缆车；在葡萄牙里斯本，大部分地区提供索道缆车服务，它们在安装了防滑装置的轨道上行驶，在城市中陡峭的坡地上不停穿梭（图13，图14）。

4.3　低收入山地住区

在许多低收入国家，平原地带通常属于富裕和中产阶级家庭，穷人们只能生活在山地上，官方的规划和开发政策常常忽视这些非正式聚落的存在。穷人们多为土地的非

Cities like San Francisco were built before people thought of reconstructing ecologies. Everybody loves San Francisco because of its steep hills: you drive a car up the steep slope and it is often not possible to see into an intersection before you reach it. Streets in San Francisco cannot exceed a 20% slope. If the street does, it is forced to become a switchback, as with Lombard Street, called the "crookest street in America." Many of the hills have been cut down to allow vehicles to reach their tops. Parking your car on a street with steep slopes is dangerous, so the direction of parking has been changed to be orthogonal. Sidewalks over 10% are often created as stairways. The great invention to deal with slopes is of course the cable car, which has become an icon for the city. Cable cars literally pull a car up the slope using a cable that's in a channel below ground (Fig. 11, Fig. 12).

Other technologies have been invented and deployed around the world to carry people, goods and services up hillsides. These include funiculars and above ground cable lifts, of which there are many varieties. In New York City, the main access to Roosevelt Island, a community in the middle of the East River, is via a suspended cable car. In Lisbon, Portugal, much of the city is served by funiculars, running on rails with gears to prevent slippage, which carry you up and down the city's steep hillsides (Fig. 13, Fig. 14).

4.3　Low Income Hillside Settlements

In many low-income countries, the flatlands are owned by the wealthy and middle class families, leaving only the hillsides for the poor. Official planning and development policies have largely neglected their informal settlements. The poor are often squatters on the land, not owning the property, and living in fear that the government will arrive one day, evict them and tear down their houses. However, in several cities in Columbia and Venezuela, there have been major governmental efforts to provide services and rationalize the development pattern of hillsides, and create the assurance that residents can continue to live there.

法占有者，因为不拥有土地所有权，所以时刻生活在恐惧之中，担心政府会在某日推倒他们的房屋并将他们驱逐出此地。然而，在哥伦比亚和委内瑞拉的一些城市，政府努力为这样的山地聚落提供服务，使山地开发趋于合理，并保证居民们可以在此继续居住。

麦德林是哥伦比亚第二大城市，坐落于河谷之中，河谷两侧的山地上则分布着一系列非正式聚落。市政府在山底的平原地区修建了一个大运量轨道交通系统，并在其中的几个站点上修建了次级交通系统，利用缆车将人们送达山顶的住所。这些坡地非常陡峭，几乎没有道路抵达，人们不得不通过疏于维护的阶梯，将大量所需物资运往位于山顶的住处，这种情况直到近年才有所改善。这些地区同样缺少安全的排水和供水系统；出于修建操场的需要，学校大多位于山下的地势平缓地区，所以孩童就学也十分困难（图15）。

对此，麦德林市政府雄心勃勃地启动了一系列开发项目，致力于改善山地居民的生活质量。其中，作为《东北城市整合计划》的组成部分，都市区缆车计划旨在利用电动缆车，将山上居民从大运量轨道交通站点送往山顶，同时围绕位于山坡上不同高度的缆车站点，集中建设面向山上居民的服务设施，包括文化机构、学校、服务中心、商店等；例如，公共图书馆允许居民选择一本图书，带回家中或附近的学校阅读。缆车站点设有顶棚，所以即使在恶劣天气条件下，人们仍然可以方便地上下搭乘。为了提高使用效率，政府还在某些地方建造了桥梁，将数个坡地连接到同一个缆车站点。同时，在低处一些不需要缆车的地方，政府安装了小型室外电动扶梯，将到达换乘站点的居民直接送往住区。

Medellín, which is the second largest city in Columbia, is located in a river valley. A series of settlements climb the steep hillsides on both sides of the valley. The city government has created a mass transit system that mainly runs on the flat lower elevations. At several mass transit stops, a secondary transit system has been built using cable cars, lifting people up the hillsides and connecting them with their houses. As the image indicates, the hillsides are quite steep, and until very recently people had to carry the bulk of what they needed up poorly maintained stairways to their homes, many of which had no road access. The areas also lacked safe sewer and water service. It was difficult for children to get to schools, mostly located on the lower elevations where the land is flat enough for playing fields (Fig. 15).

The progressive government of Medellín has embarked on an ambitious set of projects to improve the quality of life of hillside dwellers. Called *the Northeast Integral Urban Project*, one of its components is the Metrocable Project, consisting of a series of cable cars lifting residents from the mass transit stops to points high up the hillsides. Centered around the lower and upper stations on the Metrocables, there are clusters of services for the residents – cultural institutions, schools, services, shops and the like. As an example, public libraries allow people to pick up a book and take it with them to their homes or their schools nearby. Cable car stations are under cover, so that in bad weather, people can off them and on them easily. In some places, in order to be more efficient, they have created bridges to connect several hillsides to a station. And on several of the lower elevations where it doesn't make sense to run cable cars, they have installed small outdoor escalators to transport people from the transit station directly into housing areas.

图 13　纽约罗斯福岛索道电缆车
（资料来源：Greg Goodman）

Fig.13　Roosevelt Island Cable Tramway, New York
(Source: Greg Goodman)

图 14　里斯本索道缆车
（资料来源：Gray Hack）

Fig.14　Funicular, Lisbon
(Source: Gray Hack)

图 15　麦德林西裔聚居区
（资料来源：exploringkiwis.com）

Fig.15　Barrio, Medellin
(Source: exploringkiwis.com)

图 16 麦德林城市整合计划
(资料来源：CITIES Project, Cooper-Hewitt Museum)

Fig.16　Integral Urban Project, Medellin
(Source: CITIES Project, Cooper-Hewitt Museum)

图 17 麦德林电缆车入口
(资料来源：favellaissues.com)

Fig.17　Tramway Portal, Medellin
(Source: favellaissues.com)

图 18 加拉加斯西裔聚居区
(资料来源：tiendacolegial.com)

Fig.18　Barrio, Caracas
(Source: tiendacolegial.com)

总之在麦德林，政府通过实施一整套计划，切实为山地居民提供了他们所需要的交通设施、基础设施和服务设施；其中，在低处的平缓地区修建了诸如操场等需要平坦土地的设施，在高处则修建了日托所、文化娱乐等其他设施。所有这些都如同外科手术，在尽量不破坏住房和社区的前提下，以插建的方式增加各类设施（图16，图17）。

针对非法山地聚落，委内瑞拉加拉加斯市同样采取了注重整合基础设施建设和居住环境改善的策略。一方面，他们面临与麦德林相似的问题，包括缺乏排水系统、可靠的供水和电力服务以及落后的垂直交通；另一方面，他们又有自己的优势，即首期改造地区已有道路通往山顶。实际上，他们的主要问题在于半山地区缺乏向上或向下的联系。学校和操场建在高处地势相对平坦的地方，而商业功能则位于山脚；于是，不同高度之间的联系就成了改造的关键（图18）。

加拉加斯的创意是建立一个多功能的步行系统，其中部分沿用已有的步行道路，部分是新建的步行道路。每条步行道下方都建有一个基础设施层，布置水管和其他设施管道以及地面径流的排水系统等。在每个有人居住的主要台地上，都有叉出的步行道连接各处住房和其他地产（图19）。

以上两个实例表明，如果与城市开发有机结合，即便是普通的基础设施系统也能够使山地成为宜居之地，技术创新足以抵消此类地区所固有的不利条件。

4.4 山地新城

以色列的莫丁新城位于一片沟壑纵横的多岩荒漠，由摩西·萨夫迪规划而成，可谓是基于详尽山地生态分析

Thus, in Medellin, the project is a truly integrated program of adding transportation, infrastructure and facilities that hillside residents need. On the lower elevations, facilities are located that need large flat areas, such as play fields, while at upper elevations, childrens daycare, cultural activities and other attractions find their place. All of this is done in a surgical manner, inserting faciilities while minimizing disruption of homes and communities (Fig. 16, Fig. 17).

The strategy adopted by the city of Caracas, Venezuela, also focuses on coordinating infrastructure to upgrade squatters settlements on its hillsides. They faced similar issues as in Medellin, including the lack of sewer systems, dependable water supply, electtricity service and poor vertical movement. However in its initial project areas, they had the advantage of roadway access to many of the upper elevations; the problem was in the mid levels that were difficult to reach from above or below. Schools and playfields could be located on the upper elevations that were relatively flat, while commercial uses were located logically at the base of the hills. That left the connections as the main new intervention (Fig. 18).

Caracas's invention was a multi-functional walkway system. Sometimes following existing walkways, other times on new alignments, each walkway contains a layer below of infrastructure – new water and utility lines, and drainage for surface runoff. On each of the major benches where people live, there are laterals that connect to individual houses and properties (Fig. 19).

These two examples demonstrate that modest systems of infrastructure, if carefully integrated with urban development, can make hillsides decent places to live. Innovative technologies can help offset the disadvantages such areas typically bring.

4.4 Hillside New Towns

Mod'in New Town in Israel, planned by Moshe Safdie, grows out of a careful analysis of hillside ecology, in this instance a rocky desert site of ridges and valleys. It's concept is powerful but simple: recognizing that water runs downhill and is in short supply, green spaces ought to be planned for the valleys,

的结果。其概念简单有力：水源顺山而下且供应无多，所以山谷地带应规划为绿地；山坡之上景色优美，应规划为居住区域；山顶因为拥有适于建设的最佳地基条件，被保留下来进行高密度开发。由此形成一个相对厚重的生态系统，山地景观的建构也构成了这座城市的空间形态。城市的规划与自然的生态相呼应，避免了网格形态，而是强调一系列绿带汇集到市中心的大型中央公园，并在周围布置公共空间、商业空间与市政厅（图20）。

摩西·萨夫迪和他的同事并未设计新城中的任何建筑，而是基于山地生态的非常简单的设计准则，提出了具体的设计指引。其中的一条核心思想是，任何建筑的高度都不能超出树木的高度，这样在人们登山爬高的过程中，树木可以提供阴凉空间，创造宜人的环境氛围。新城中的土地被分割后出售给个体开发商，由其各自邀请建筑师进行建筑设计，但整体的规划结构依然清晰可见（图21，图22）。

山脚下的线型绿色空间郁郁葱葱，种满了不需要每天浇灌的沙漠植物。一条阶梯状的街道使行人得以从一座主要的山上缓步下行，在沿街道分布的商业区里停留购物，并且通过绿带将各处的高密度开发联系在一起。

最后一个例子是重庆的悦来生态城，由彼得·卡尔索普负责设计，是在中国南方特定环境中进行高密度山地开发的尝试。这个项目在"生态"方面的设计可谓中规中矩，但在尊重山地建设需要方面的设计十分精彩，提出了一个可持续的生态系统（图23）。

悦来生态城位于山顶，通过一座新建的桥梁与外界相连。基于仔细的土地分析，最为陡峭的坡地被划定为保护区域，禁止进行开发建设。尽管卡尔索普因为倡导网格化道路

and housing ought to occupy the hillsides where views were possible. This results in a relatively thick ecology. The tops of hills were reserved for some of the highest densities because they had some of the best foundation for development. The morphology of that city consists of a constructed hillside landscape. The plan of the city is responsive to the natural ecology, eschewing the grid in favor of a series of green ribbons that merge in a large central park in the center of the city, with the public spaces, commercial spaces and city hall nearby (Fig. 20).

Moshe Safdie and his colleagues designed none of the houses or buildings in this town. Instead, they produced design guidelines, based on a very simple set of design rules that prescribe the hillside ecology. One key idea is that nothing should be taller than the height of trees so that as you make your way up the hillside the trees provide shade for these surfaces creating a hospitable environment. Each parcel of land was sold to an individual developer, that hired its own architects to design structures. But the general scheme is clearly evident from the illustrations (Fig. 21, Fig. 22).

The linear green spaces at the base of the hills are quite lush, lined with desert plants that don't require irrigation every day. A great cascading street allows pedestrians to make their way down one major hill in gradual steps, stopping, patronizing commercial outlets distributed along the walkway, connecting high density development with the green ribbons.

My final example is Yuelai Eco-City in Chongqing. Designed by Peter Calthorpe, this is an attempt to create a high-density hillside development in the particular environment of South China. While the "Eco" components are fairly standard, its plans do a good job of respecting the demands of hillside construction, proposing an ecology that is sustainable (Fig. 23).

Yuelai is located mostly on the top of a hill, reached by a new bridge crossing. Based on a careful land analysis, the steepest slopes have been identified and reserved from development. Although Calthorpe is known for his advocacy of grid plans, the topography here requires that roadways be distorted to fit the topography. Three shelfs of development are proposed; the problem is how to connect them. The upper levels are largely connected by mass transit, below ground.

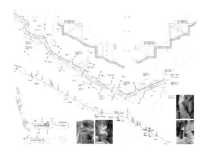

图19 加拉加斯多功能步行系统
（资料来源：CITIES Project, Cooper-Hewitt Museum）

Fig.19 Multi-Functional Walkway System, Caracas
(Source: CITIES Project, Cooper-Hewitt Museum)

图20 以色列莫丁新城
（资料来源：Moshe Safdie Association）

Fig.20 Mod'in New Town, Israel
(Source: Moshe Safdie Association)

图21 莫丁新城的绿带
（资料来源：Moshe Safdie Association）

Fig.21 Green Ribbon, Mod'in
(Source: Moshe Safdie Association)

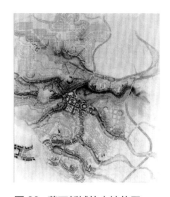

图22 莫丁新城的山地住区
（资料来源：Moshe Safdie Association）

Fig.22 Hillside Housing, Mod'in
(Source: Moshe Safdie Association)

图23 重庆悦来生态城
（资料来源：Calthorpe Association）

Fig.23 Yue Lai Eco-City, Chongqing
(Source: Calthorpe Association)

图24 悦来生态城扶梯的布局
（资料来源：Calthorpe Association）

Fig.24 Location of Escalators, Yue Lai Eco-City
(Source: Calthorpe Association)

而闻名，但这里的地貌显然要求对道路形态进行调整以适应地形的变化。规划提出在三个高度层面上进行开发建设，由此而来的问题便是如何将三者联系在一起。最终，上部的开发主要通过地下大运量轨道交通系统与外部联系；位于山坡上的主要开发廊道则通过电动扶梯系统与大运量轨道交通系统相连，沿线布置公共服务设施；最陡峭的坡地被保留下来禁止开发，未来将成为重要的排水通道和新的社区公园，形成一个处处连通的开放空间系统（图24）。

4.5 结论

在结束时，我要重审文章开篇提到的观点：针对一个场地，仅仅保护它的自然系统是远远不够的，占用山地会不可避免地改变这些生态系统。所以，我们的任务是构建一个可持续的居住生态。这意味着要构想一种不同的生活模式，意味着寻找使得人流垂直运动的方法，使用新的交通运输设施；意味着创造面向所有居民开放且便于通达的开放空间，使人们可以享受山地带来的各种好处，其间可能需要引入一些与山地住区全新的微气候环境相匹配的景观要素；也意味着拥抱山地的无限潜能，而不仅仅是解决问题。面对山地，你必须思考与平原环境截然不同的生活模式，最优秀的山地城市一定对地貌、生态和生活模式做出了独特的回应。

The major hillside corridors will be connected to the mass transit systme by a series of escalator systems, lined with public facilities and services. The steepest slopes will be reserved from development, as will the main drainage courses. These will become the new community parks, a connected system of open spaces (Fig. 24).

4.5 Conclusions

I end where I started: that it is not enough to preserve the natural systems on a site. Occupyiing a hillside inevitably changes those systems. Rather, the task is constructing an inhabited ecology that is sustainable. This means conceptualizing a different pattern of life. It means, finding ways by which people can move vertically, walking and using new transit devices. It means creating open spaces that are accessible and open to inhabitants, and allow them to enjoy the benefits of being on the hillsides. It may involve importing landscape materials that recognize the new microclimates of the inhabited hillsides. It means celebrating the possibilties of the hillside, not simply mastering the difficulties. You have to think differently about the pattern of life on a hillside than you do in a flatland environment. The best hillside cities are unique responses to topography, ecology and living patterns.

参考文献 / References

CINTHIA E S, 2007. CITIES: Design with the other 90%[M]. Cooper-Hewitt: National Design Museum.

HACK G, 2018. Site Planning: International practice[M]. Cambridge: MIT Press.

HORST J S, HACK G, 2007. Landforming: An environmental approach to hillside development, mine reclamation and watershed reclamation[M]. Hoboken: John Wiley & Sons.

赵光耀，2008. 山坡地生态稳定与经济持续发展技术研究[M]. 郑州：黄河水利出版社.

1950 年以来的中国城市演进之路
URBAN FORMATION IN CHINA SINCE 1950 SEEN FROM AFAR

彼得·罗
哈佛大学杰出贡献教授
哈佛大学设计学院教授
清华大学建筑学院荣誉教授

Peter G. Rowe
Distinguished Service Professor, Harvard University
Raymond Garbe Professor of Architecture and Urban Design, Graduate School of Design, Harvard University
Honorary Professor, School of Architecture, Tsinghua University

---

中文翻译：刘佳燕，张琪
Chinese translation by Jiayan Liu, Qi Zhang

谈及今天的中国，我希望从一系列宽泛的国家特征开始，探讨三个方面的问题。初窥之下，中国被划分为34个省级单位和1679个县，两者都是长期使用的行政区划单位。中国领土范围内地势和地形的差异相当大，从山地到海滨，资源分布不均。同样是水资源，在西部和北部地区极为稀缺，在东部和南部地区则相对丰富。1950—2010年的人口增长一直是不规律的，特别是在毛泽东时代、婴儿潮和生育低谷期、经济反复无常的时期，以及如1960年代初大饥荒的年代。值得一提的是，中国人口总数从1950年的5亿，经历50多年上升到今天的超过13亿。这与美国的情况也有些类似，1950年美国人口总数为1.52亿，当前为3.3亿，人口规模翻了一番。

可以肯定的是，中国人口的空间分布是不均衡的，县级行政区的人口密度某种程度上反映了地形、地貌以及其他自然因素的空间分布。显然，中国的人口集中在沿海和中部地区。另一种更为精确度量人口分布的方法是统计县级行政区的住房密度。虽然数据资料是从2000年开始的，但其结果显示却大致相同。城市和主要镇的分布也表现出类似的趋势，同时在空间上具有很强的集聚性，从北到南集中于：（1）环渤海地区，包括北京和天津；（2）山东省；（3）长江三角洲地区，包括上海；（4）珠江三角洲地区，包括广州。这些集聚区基本上都位于三大流域。

当前中国的城市聚落建制主要分为六类。此外还有"建制镇"这种类型，指非农业人口超过2000人，同时具有一定的密度和其他城市特征的地方。建制镇的平均人口规模约8500人。在工作中通常还会用到另外三种类型：（1）城区，包括农业和非农业人口，以及周边居民；（2）市区，范围大致与各城区外边界相吻

In talking about China today, I'm going to tell three stories and I would like to begin by looking at some broad national characteristics. At first blush, China is subdivided into 34 provinces and 1,679 counties, both of which are longstanding. The topography and terrain varies considerably, from mountains to coastal areas, as do matters of resource distribution, such as water, which ranges from scarce in the north and west, to relatively abundant in the south and east. Population growth from 1950 to 2010 has been irregular, especially during the Maoist era, through "baby booms and busts", uncertain economic times, and the famine in the early 1960s. Notably, the population rose from 500 million in 1950 to above 1.3 billion today in a little over 50 years. This, however, is somewhat comparable to the U.S. which also doubled from around 152 million in 1950 to 330 million inhabitants today.

Spatial distribution of China's population is uneven to be sure, returning to enumerations of density by county, somewhat mirroring topography, terrain, and other natural factors. Clearly, it is loaded to the coastal and central areas. A sharper measure of population distribution is provided by housing density in counties. Although dated from 2000, the result is much the same. The distribution of cities and major towns shows a similar tendency, but also with strong agglomerations, moving from north to south around: (1) the Bohai Bay Region, including Beijing and Tianjin; (2) Shandong Province; (3) the Lower Changjiang Delta, including Shanghai; and (4) the Pearl River Delta, including Guangzhou. These essentially are all located within the three major river basins.

Today in China, designation of urban settlements falls into six classes. To these can be added "designated towns", ie., places with non-agricultural populations above 2,000 people, as well as appropriate densities and other urban-defining characteristics. Across all the designated towns, the average population is about 8,500 people. Also at work are three other classificatory distinctions: (1) urban districts, including non-agricultural and agricultural populations, as well as peri-peripheral dwellers; (2) regional-level cities, more or less coterminous with urban districts but accounted for in the non-agricultural populations; and (3) county-level cities, accounted for in the non-agricultural populations. In other words, a distinction can be made between

合，但主要以其中的非农业人口为评价标准；（3）县级市，主要以非农业人口作为评价标准。换句话说，这样可以区分"大家伙"——市区和城区和"小家伙"——县级市和建制镇。中国共有657个地级市，还有约19500个建制镇。

中国的城市化进程自1950年以来的前三十年间一直进展缓慢，这亦是毛泽东时代"非城市"偏好的结果。进一步从时间上来讲，1979年，城市人口占总人口的比例只有17.9%。此后经历了五到六个不同的发展阶段，城市化水平到今天已接近50%。自1978年以来，约有3.5亿人进入城市地区，这一增长主要通过人口从乡村向城市以及在城市之间的迁移完成。

从空间而言，许多大城市地区的发展一直相对较慢，只是在近期出现了急剧扩张。1949年，中国仅有为数不多的几个大城市，人口主要分布在农村。到了1957年，人口快速增长，但具有一定规模的城市数量几乎没有增加。1965年，由于1962—1963年的饥荒，经济情况相当差。即使到1978年，作为邓小平"改革时期"的开局年，这种情形也没有太大的变化。1985年城市化开始加速，到了1990年，中国开始经历急剧加速的城市化进程。从2005年到现在，具有一定规模的城镇的分布越来越密集。

1978年以后的城市扩张主要来自以下几个方面的驱动。首先，社会主义市场经济制度的建立和1994年的财政改革推动中国的经济环境得到极大改善。一个直观体现是全国人均GDP达到了8000美元左右。参考国际经验，在此经济发展阶段将形成大规模的中产阶层。经济情况的改善亦是由于经济结构的调整，遵循标准的发展轨迹，经

the "big guys" – regional-level cities and urban districts, and the "little guys" – county-level cities and designated towns. In total, there are around 657 designated cities, plus about 19,500 towns in China.

During the first 30 years, or Maoist period with its non-urban bias, urbanization has been slow in coming to China since 1950. Further on in time, in 1979, the urban proportion of the total population was only 17.9%. Then, through some five or six phases, with varying rates, urbanization today approaches 50% of the total population. Since 1978, about 350 million people have found themselves in urban circumstances. Primarily, this increase has come about through rural-urban and urban-urban migration.

Spatially, the distribution of numerous major urban areas has been relatively slow to materialize also, although more recently with more dramatic expansion. In 1949, there were few large cities, the distribution of the population was primarily rural. In 1957, there was rapid population growth but few additional sizeable cities. In 1965, there were poorer economic conditions, because of the famine of 1962–1963. By contrast, 1978 was the beginning of the "reform period" under Deng Xiaoping, but there was not much change. In 1985, there was some acceleration in urbanization, and by 1990, China saw the beginning of rapid acceleration in urbanization. In 2005, there was a much denser distribution of sizeable cities and towns, continuing on into the present.

The major drivers of recent expansion, post-1978, are several-fold. First, China saw vastly improved economic circumstances for many, under the rubric of a socialist market economy and the 1994 fiscal reforms. This improvement amounts to a level of around $8,000 per capita, approaching the economic point of substantial middle-class formation elsewhere, as well as in China. Economic improvement was also due to sectoral shifts within the economy, roughly following a standard "development trajectory", from primary, secondary, to tertiary. Nevertheless, if anything, services still lag, with big gains still in industry, often export-oriented. By contrast, the labor component of this segmentation is very different, still with a large agricultural labor pool. Substantial investment in public infrastructure, like roads and especially inter-regional and inter-

济重心从一产、二产，逐步向三产转移。不过，服务业的发展仍然滞后，工业依旧占据主要地位，而且往往是出口导向型。与此相反，劳动力的组成却与上述生产结构呈现巨大差异，农村劳动力一直占据很大比重。投资主要集中于公共基础设施建设，例如道路，特别是跨区域、跨城市的道路，为各个生产中心之间提供必需的联结，以促进发展。这在目前仍是一个未完成的使命。与之类似的，主要的铁路线路得到了进一步的发展，为客运和货运提供了更好的服务。尤其值得一提的是，当今雄心勃勃的全国高速铁路网络建设，以"四横四纵"为骨架便捷串联起北京、广州、武汉、上海、杭州、天津和其他诸多城市。

由于家庭规模的缩小和住房面积标准的不断提高，在两者的乘数效应影响下，城市地区呈现加速扩张的态势。使用以下三种统计方法都显示人均住宅面积已显著提升：（1）"居住空间"，指排除厨房、卫生间、服务用房、走廊和剩余空间以外的居住性空间；（2）"使用空间"，包括除剩余空间和死角外的所有空间；（3）"建筑空间"，指住宅的总建筑面积。值得指出的是，同其他地区一样，中国农村的平均居住水平通常高于城市。现在的普遍状况是人均住宅建设面积为20~25平方米，而到2025年或2030年，人均建设面积将上升至32~35平方米。目前的一个改革趋势是供应更多的可支付性（面积较小的）居住单位，比如"70-90"政策要求70%的住宅建筑单元面积不得超过90平方米。家庭规模下降和空间指标提升这两大趋势的"乘数效应"带来了对城市居住空间的更多需求，未来还将持续增长。

此外，还有其他一些重要的城市化现象，比如乡村工业化，以及建制镇和建制乡的城市化，特别是自1978年

city roads provided needed linkage among productive centers, boosting development. This is also an unfinished project. Similarly, major rail links have been further developed, also providing improved access to both passengers and freight. Indeed, this is especially the case with today's ambitious high-speed rail network, with its four vertical axes and four horizontal routes, readily linking Beijing, Guangzhou, Wuhan, Shanghai, Hangzhou, Tianjin, and numerous other cities.

Then too, there has been the accelerated expansion of urban areas under the multiplier effect of shrinking household size and rising housing space standards. Residential space standards per capita have risen appreciably across all three statistical measures that are used: (1) "livable space standards", which refers to the exclusion of the kitchen, bath, services, and corridor or residual space; (2) "useable space", which is inclusive of all except the residual or dead space; and (3) "built space", which refers to the built residential gross floor area. Notably, rural standards are often higher on average than urban, as elsewhere. Generally now, 20 to 25 m² per person is being built consistently, with likely rises to around 32 or 35 square meters per person by 2025 or 2030. There have been corrections in the direction of providing affordable (smaller) housing units, however, with for instance the "70-90" rule, whereby 70% of residential construction must be equal to or no more than 90 m², plus other provisions. The nominal 'multiplier effect' of both trends – declining household size and increased space standards – boosts demand for urban residential space, also rising into the future.

In addition, there are other significant urbanizing phenomena like the rural industrialization and urbanization of designated towns and townships, especially during the immediate post-1978 period. This gave rise to the Township Village Enterprises (TVEs) and spontaneous in-situ industrialization and urbanization, linked to excess labor from the agricultural revolution that took place in the early 1970s. By 1990, if not before, spatial distributions in the form of "Desakota regions"or "urban continua" began to establish, especially in the peri-periphery and countryside of the major river deltas. In fact, production there accounted to fully 34% of national GDP in 1993, before tapering off to around 5%-6% today. Nevertheless,

至今的这段时期。1970年代初的农业革命解放了大量剩余劳动力，促成了乡镇企业的发展和自发的就地工业化和城市化。到1990年，"城乡结合部"和"城市连绵带"的空间形态开始出现，尤其在几个主要河流三角洲地区的城郊和乡村地区。事实上，1993年这些区域的生产总值曾经高达国内生产总值的34%，而今天这一比例已逐步降至5%～6%。然而，作为一个整体，全国县级市总的生产总值仍然占到国内生产总值的约56%；在农村地区，家庭收入中依靠非农就业获得的部分占比至少达到50%。

此外，"城中村"现象已经出现，尤其在中国南部以及广东省及其周边地区，西至昆明市（图1）。城乡二元土地制度下，村集体主要负责集体土地所有权，从而主导整个村庄的发展。很多土地都被用于房屋出租业务，特别是为涌入城市的流动人口提供住房。结果是一批村民成为以收租金为生的"食租者"。如果考虑户籍状况，估计全国范围内流动人口的总规模有1亿，或者略少些。

今天，同过去一样，许多大城市发现自己仍处在被村庄和农田紧密包围的环境之中。这在北京确是事实，因其拥有广阔的大都市区域。在885平方千米的地区中大约有13000个村级定居点。它们的规模差别很大，平均约300人，平均面积约30公顷，空间分布上主要在四环路以外的地区。许多村庄在自然灾害预防和自然保护方面存在风险或面临负面影响，重大公共安全和福利问题也日益凸显。其中，自然灾害影响区包括易滑坡的地区、地震易发区和洪泛平原；自然保护区包括指定的绿带、绿地保护区和历史遗迹区。

当前在建设"和谐社会"的战略指引下，乡村问题再次备受关注，伴随以对乡村定居点进行合理调整和发

the smaller county-level cities in the country as a whole continue to account for around 56% of national GDP, and with non-agricultural employment in China's countryside accounting for at least 50% of household income.

In addition, an "urban village" phenomenon has emerged, particularly in southern China in and around Guangdong Province and as far west as Kunming (Fig. 1). Under the dual system of land tenure, rural collectives operating largely with indivisible property rights hold sway over villages. Moreover, many have gone into the rental housing business among other enterprises, particularly for members of the floating populations that have swarmed into many cities. The result is a rentier situation on the part of the villagers that can often be exploitative. Estimates of the aggregate size of the floating population in China range from over 100 million to somewhat less; defined by way of exception to the hukou or household registration system.

Today, as in the past, many large cities still find themselves in a relative "dense-pack" of village and rural circumstances. This is certainly true of Beijing within its extensive Greater Metropolitan Area. There are around 13,000 village-level settlements in an area of some 885 km^2. Sizes are relatively variable, with an average population of 300 people, and average areas of around 30 hm^2. Spatial distribution is largely well outside of the 4th Ring Road. With respect to natural hazards and nature preserves, numerous villages are at risk and adversely affected, raising significant public safety and welfare issues. Here, hazards include: landslide areas, earthquake areas, and flood plains; preserves include: designated greenbelts, green land preserves, and historic sites.

Certainly under the present re-emphases of rural circumstances under the rubric of a "harmonious society", these conditions have led to strategies for rational rearrangement and reinforcement of village settlement, of which there are three levels: (1) spatial re-ordering, including modifying layouts, improving services, etc; (2) relocation and resettlement, including moving populations, having fewer villages, and getting people out of harm's way; and (3) maintenance and development of villages in place. The first and second in sum would constitute nearly 50% of the total, with the third type

图1 昆明城中村里的居民
（资料来源：Peter G. Rowe）

Fig.1 Urban villagers in Kunming
(Source: Peter G. Rowe)

展的相关政策，并主要体现在以下三个层面：（1）空间的重组，包括调整布局、改善服务等；（2）移民的搬迁和安置，包括转移人口、缩减村落，以及将人口迁离危险地区；（3）村庄的就地保留和发展。第一类和第二类政策涉及村庄占总数的近50％，第三类政策涉及余下的50％。类似的策略在一些地方也得以发展和实施，比如拥有约30万村民的上海浦东地区。

回到宏观层面，上述事件的结果再加上其他城市化进程的拉动，使得到了2005年，人口的空间分布集中在东部沿海，而向国家中部转移的趋势开始减缓。人口聚集区不出所料地位于二产和三产发达地区。反过来，这些集聚区又青睐拥有内在竞争优势、规模经济和较好市场可达性的地方。各种新兴模式都或多或少地遵从于传统的或标准的现代化发展途径。自1986年城市建制进入稳定发展阶段以后，数据开始具有可比性，城市数量从324个增长到657个，并在1997年达到688个的顶峰值。之后主要由于行政区划合并，城市数量有下降的趋势（图2）。

1978—2002年，城市规模的发展呈现以下趋势：（1）超大城市和特大城市（非农人口大于100万）的数量从1985年的17个上升至1990年的31个，如今超过了45个；（2）中等城市（人口在20万～50万）数量占比在1995年之前呈小幅增长，之后变化不大；（3）大城市数量缓慢上升；（4）小城市数量占比上下波动，在1995—1996年间达到了峰值。2005—2006年间的数据统计也显示了大致相同的趋势。不过，超大城市和特大城市如今展现出发展优势，按非农业人口数量统计有52个，若按城区统计则达到了113个。大城市数量的占比也在不断攀升。这两种走向都说明了集聚趋势和规模经济的影响。而

constituting the remaining 50%. Similar strategies are being development and implemented elsewhere, such as in Shanghai and Pudong which has around 300,000 villagers.

Returning to the "bigger picture", the result of these, plus other urbanizing "pulls" yields a spatial distribution for 2005 with strong concentrations on the East coast, lessening as one moves to the center of the country. Not unexpectedly, the distribution shows agglomerations in major areas of secondary and tertiary production. These agglomerations, in turn, favor locations with inherent competitive advantages, economies of scale, and access to markets. The emerging pattern, more or less, conforms to a conventional or standard pathway forward in modernization. Since 1986, when urban designations became reasonably stable and, hence, comparable, the number of cities has expanded from 324 to 657. It peaked in 1997 with 688, followed by a downward trend due to jurisdictional consolidations, mainly (Fig. 2).

Seen from the perspective of city size between 1978 and at least 2002, there has been: (1) a rise in the number of mega- and super-size cities (> 1 million people) in non-agricultural population from 17 in 1985 to around 31 in 1990, and over 45 today; (2) some increase in the medium-size class (200,000 to 500,000 people) as a proportion of the total, before flattening out beyond 1995; (3) slower rise in the large-size class; and (4) fluctuations in the small-size class, which also peaked in proportion of the total around 1995–1996. A snapshot in 2005–2006 shows much the same trend. However, now mega- and super-large cities are in the ascendency at 52 in number from non-agricultural population and 113 as urban districts. The large-city class's proportion is also rising, with both trends suggesting agglomeration and economies of scale. While the medium-size city class and the small-city class are claiming less of the overall population, i.e., an overall a push upward in scale or towards a more conventional pathway forward.

Outside the "city classification", designated towns or townships grew at a faster rate in population than the city populations at least since 1995 – the TVEs early success. More recently, however, the trend lines have leveled off with cities now in the ascendency. But, overall, the presence of a bi-polar distribution of urban settlement is suggested in the direction of

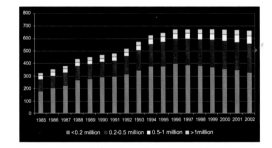

图2　1985—2002年间不同规模城市的数量变化
（资料来源：中国统计年鉴）

Fig.2　Changes in the numbers of different sized cities, 1985—2002
(Source: China Statistical Yearbooks, various)

中等城市和小城市中的人口数量则呈现缩减。整体而言，城市规模呈上升趋势，或者说中国城市正迈向一条更常规的发展道路。

在"城市分类"之外，至少自1995年以来，建制镇、建制乡由于乡镇企业在早期的成功，人口规模增长速度甚至超过了城市。不过，最近乡镇的人口规模增长已趋于平缓，而城市正处于优势地位。但总体而言，城市人口规模的分布呈现两极分化，预示着"大家伙"和"小家伙"共存的发展方向（图3）。这并非特例。例如在意大利，这个被称为"由100个城市组成的国家"拥有大量的小型社区，所有城市中只有4个城市的人口数超过了100万。

大体上讲，自1978年以来，中国城镇的分布变化似乎遵循着国家政策的安排，并可分为四个主要阶段。第一，1980年的"全国城市规划工作会议"提出要控制大城市，合理发展中等城市，积极鼓励小城镇的发展。第二，1989年颁布的《城市规划法》和其他的刺激性政策给予直辖市自主制定规划的权力，1994年的财税体制改革推动出让"土地使用权"作为地方财政收入的来源，这些带来了1995—1997年间"城邦"的崛起。第三，2001年的第十个五年计划以及加入WTO以后，国家开始限制大城市任意发展，或者说要求至少其发展必须是合理的。合理开发的要求同时也面向小城市提出，要求提高效率和减少污染。第四，第十一个五年计划中再次包含了针对特大城市和大城市的内容，强调的重心转移到功能多样化和促进乡村发展，同时鼓励合理开发二、三线城市。

比较显示，上海等超大城市地区的人口规模低于美国及其他发达地区的水平。同时，相比于对数正态分布曲线而言，城市规模分布曲线在小规模一端"向外隆起"，再

the "big guys" and the "little guys" (Fig. 3). This is not unusual if we recall small communities, such as in Italy, which is also referred to as the "country of 100 cities", with only four of them with populations at 1 million or greater.

By and large, over the period since 1978, the distribution of cities and towns appears to have followed national policy of which there have been perhaps four major stanzas. First, the National Conference on Urban Planning in 1980, which called for the control and containment of larger cities, the rational development of medium-sized cities, as well as the vigorous encouragement of growth in smaller cities and towns. Second, the *City Planning Act* of 1989 and other stimuli, including giving municipalities the right to create plans along with the 1994 fiscal reforms that included the sale of "use rights" as revenue streams for municipalities. This marked the beginning of the rise of the "municipal state", around 1995–1997. Third, the 10th Five-Year Plan of 2001 and the post-WTO period, sanctioning larger cities to develop at will, or at least rationally. Rational development was also pressed upon smaller cities to be more efficient and less polluting. Fourth, the 11th Five-Year Plan in which super-large and large-cities were to be contained again, shifting the emphasis to functional diversification and rural development, while rational development was encouraged for 2nd and 3rd-tier cities.

One consequence is that Shanghai and other mega-cities are smaller in population than might otherwise be expected in a well-developed hierarchical pattern of settlement to be found in, say, the U.S. and other parts of the well-developed world. Also, substantial interdependencies are suggested by the "outward bulge" in the lognormal distribution of city size at the smaller end of the spectrum, returning again to the bi-polar distribution of urban settlement. Also, this coincides with some geographer's observations that China is under-urbanized given the level of economic development.

Looking at this phenomenon more closely, we see comparatively dissimilar distributions between the U.S. and China, as well as a time series of distributions for the Changjiang Delta. These distributions also suggest that circumstances are becoming more "normalized" or "conventional" as far as alignments of scale and rank are concerned, with the upward

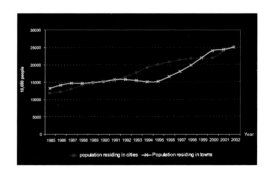

图3 1985—2002年间城市和建制镇中居住人口的规模变化
（资料来源：中国统计年鉴）

Fig.3 Changes in the numbers of different sized cities, 1985—2002
(Source: China Statistical Yearbooks, various)

次印证了城市聚落的两极分布。这也正好与一些地理学家的观测一致,即中国尚未达到与经济发展水平相一致的城市化率。

仔细观察这种情况,我们可以发现美国和中国在城市规模分布上的不同特征,以及长江三角洲地区城市规模分布随时间推移的变化情况。研究显示,城市规模及其排名情况正在变得日益"标准化"和"常规化",每年的分布曲线都会向上移动(图4)。如果研究对象换成特大城市(人口超过100万),也会得到类似的分布图。除了特大规模的城市数超过美国外,中国大城市的数目相对较少。不过,大城市的比例正在逐步上升。

我们再来看国家层面的统计数据,我从中总结了七个方面。

第一,城市发展的区域失衡现象日益严重,非农人口的分布从1978年沿海、中部和西部地区各占1/3的状态,演变成现在的沿海占44%,中部38%,西部18%。这个数量的变化相当于向沿海地区净转移了1.6亿的人口,向中部净转移了6600万人口,相应地西部地区则净流失了同等规模的人口。总体而言,这是由经济发展重心向东部沿海地区转移及其拥有的比较优势所导致的。各省人均GDP分布图也在一定程度上反映了这一状态。

针对这种失衡现象,第十一个和第十二个五年规划中特别提出了要实现"五个平衡":(1)平衡社会服务发展与GDP增长的关系;(2)平衡东西部地区之间的差距;(3)城乡平衡和农村生活改善;(4)协调人与环境的关系,重视可持续发展;(5)强调国内与国际市场之间的协调发展,平衡生产和消费。此外,尤其需要关注内陆城市,如重庆、武汉、贵阳、昆明等地区的发展,重

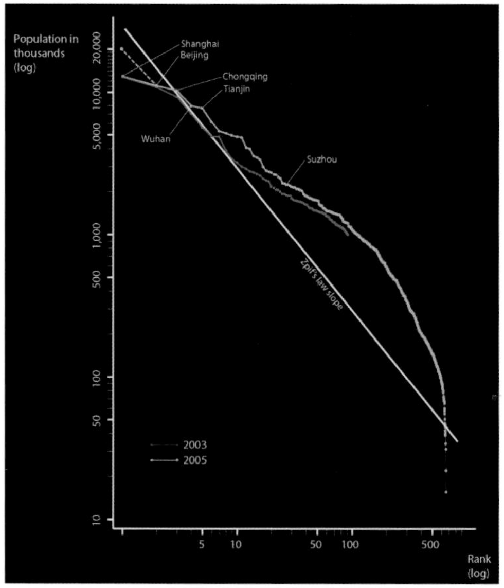

图4 长江三角洲地区城镇的弗兰克规模分布
(资料来源:Peter G. Rowe)

Fig.4 Frank-size distribution of cities and towns in the Changjiang Delta
(Source: Peter G. Rowe)

rise of yearly distributions (Fig. 4). Then too, these distributions make similar points in a different way by looking at big cities, ie., populations of more than 1 million people. China has relatively fewer big cities, except at the extremes of size than the U.S. The proportion of large cities, however, is rising in China.

Now, looking over these and other national-level statistics, several observations can be made of which I will expressly enumerate seven.

First, there has been a growing imbalance regionally in urban development from roughly parity 1/3-1/3-1/3 to coast, center, and west, for non-agricultural populations in 1978 to 44% coastal, 38% central, and 18% western today. This amounts to a net change to the coast of 160 million people, and to the center of 66 million people, together with a net loss of about the same magnitude in the west. Largely, this is driven by economic development emphases and comparative advantages of the coastal area. In part, this shows in the map of provincial GDP per capita.

This imbalance has received attention under the 11th Five-Year Plan and the 12th Five-Year Plan with emphases on the "5 Balances": (1) even development – social services vs. GDP growth emphasis; (2) leveling East-West disparities explicitly; (3) rural-urban and improvement of rural life; (4) better man-environment relations, ie., sustainability; and (5) emphasis on domestic vs. international and moving away from the skewing of production and consumption. Also, substantial emphasis on inland cities like Chongqing, Wuhan, Guiyang, Kunming, etc, with regards to funding and leadership have taken place. Parenthetically, geographic spatial imbalances are not unprecedented elsewhere. For instance, the U.S. can be regarded as something of a "lopsided donut", with relatively less population towards the west of its geographic interior.

A second observation is that China has relatively few very large cities by international comparisons. We already saw this with regard to the U.S. where there are 50 cities of more than 1 million in population while in China, 52 with non-agricultural population counted and 113 with urban districts population counted. The largest city is Shanghai, which is smaller than Tokyo, Mumbai, and Mexico City (Fig. 5). Also, the proportion of the total population living in one or several very large cities

视资金投入和带动效应。这种地理空间的失衡并非中国独有。例如，美国可以被视为一个"倾斜的多纳圈"，西部地区的人口相对较少。

第二，在国际比较的视角下，中国的特大城市数量还相对较少。在美国，人口超过100万的城市有50个，而在中国，按非农人口计算这样的特大城市有52个，按城区人口计算有113个。上海作为人口规模最大的城市，其人口总数仍然不及东京、孟买和墨西哥（图5）。此外，特大城市中居住人口占总人口的比例仍然很低，仅为18%。东京、大阪和名古屋3个城市的人口数占日本总人口的45%，首尔的人口数占韩国总人口的20%，整个首都市圈的人口则占到49%，布宜诺斯艾利斯的人口数占阿根廷总人口的32%。

第三，尽管中国城市化人口迁移的绝对量级在历史上是前所未有的，但放眼全球，到目前为止并非独一无二。美国的"大洗牌"历程带来了5000万移民，约占战后总人口的30%，其规模大致相当于中国目前的迁移人口。类似的情况也发生在战后的意大利，移民人口占比25%~27%。毫无疑问，如果中国延续目前的发展轨迹，很可能将拥有世界上最大规模的城市转移人口。不过，截至今日尚未达到。

然而，当我们转向现代化建设中的另一个发展和建设投资的指标建设规模，中国的情况可以说是史无前例的。2007年，中国的建设支出占全球总量的42.5%，即接近一半（图6）。此外，这一比例还在不断升高，年均增长速度是世界其他地区的1.7~2.0倍（印度除外），远远超过了城市人口的增速。不过需要注意的是，这里说的建设不单单指城市建设，而是包括一切相关的建设。例如，住

is small at 18%. Tokyo, Osaka, and Nagoya constitute 45% in Japan, Seoul at 20% but the metropolitan area at 49% for the whole of Korea; while Buenos Aires constitutes 32% of the population of Argentina.

A third observation, is that while the absolute magnitude of urban population migrations in China is historically unprecedented, in normalized terms, so far it is not. The "Great Reshuffling" in the U.S. embraced 50 million people, approximately 30% in the post-War era, and roughly equivalent to what has occurred in China. The same happened in post-War Italy at a little less with 25% – 27%. No doubt, as China continues its current trajectory, it may well surpass the normalized scale of urban transition of other places. So far, however, it has not.

However, turning to another indicator of development and capital investment in modernization like the volume of construction, China may well have moved into unprecedented territory. In 2007, based on construction spending, China accounted for a massive 42.5% of the global total, ie., approximately half of the total (Fig. 6). Moreover, if anything this proportion is increasing with an average annual growth rate of 1.7 to 2.0 times other places in the world, with the possible exception of India, and well ahead of the urban population growth rate. One should note, however, that total construction includes everything and not only urban construction. For instance, residential construction constitutes almost 10% of annual totals.

It is probably fair to say that the rate of urban growth by population is not yet unprecedented but the rate of urban building transformation may be largely unprecedented. Contributing factors here are: (1) increases in residential and other space standards; (2) shifts to service industries; (3) urban infrastructure improvements. Nevertheless, yet again, this last observation may also not be entirely the case. Numbers for the U.S. in its post-War boom, including the construction of some 55,000 miles of highway, accounted for a very similar amount of the world's total construction and with a much smaller population base.

A fourth observation is that average urban growth rates in China, although reasonably high are not very high, nor extremely

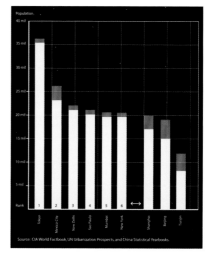

图5 主要城市的人口排名
（资料来源：World Cities Population, *Demographica*, 2012）

Fig.5 Ranking of major city populations
(Source: World Cities Population, *Demographica*, 2012)

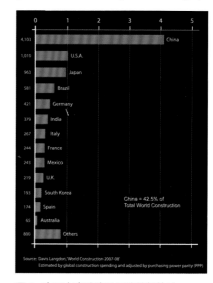

图6 中国占全球建设量的份额估计
（资料来源：Davis Langdon, World *Construction*.）

Fig.6 Estimate of China's share of global construction
(Source: Davis Langdon, World *Construction*.)

宅建设量占全年总量的近10%。

所以一种更为精确的说法是，中国的城市人口增长速度尚未达到前所未有的水平，但其城市建设转型的速度却是空前的。贡献因素包括：（1）居住和其他空间标准的提高；（2）经济重心向服务行业的转移；（3）城市基础设施的改善。不过，这一情况也许同样并非独一无二。美国在战后繁荣时期的大规模建设，包括88000千米的高速公路建设，在当时世界建设总量中占到了相似的比重，但却是基于一个比中国更小的人口基数。

第四，中国城市平均增长率是相当高的，但并非很高，更算不上极高。实际上，它低于首尔、东京（图7）。这也可能是使得中国城市幸免于淹没在棚户区和贫民窟海洋中的原因之一。

第五，中国的城市化模式呈现显著的两极分化，并且伴随位于两端城市的快速发展，这一趋势可能继续强化。进而言之，这种两极分化可能会令中国在面对社会和环境问题时尝试寻求差异化的解决方式，在不同层级提供良好的、多元化的服务和居住空间，而不仅仅依赖于大城市。

第六，中国的城市化和工业化发展已经远远超出环境承载能力，经济生产的影响超越了环境的修复能力。根据最近对"环境弹性"的测算，例如在环境变化与经济变化的关联关系方面，中国得分—0.43，意味着1%的经济增长伴随以环境质量下降0.43%，可持续性较低，对比数值是荷兰0.04，日本—0.01，美国—0.13。中国科学院最近的一项研究显示，环境修复成本将导致原本为9%～10%的GDP增长值下降4%。

第七，中国拥有较高的居住密度，但总体而言，城市地区的居住密度正在下降。从全国来看，657个城市平均

high. They are certainly below growth rates experienced over smaller periods of time for Seoul, Tokyo, (Fig. 7). Also, this is probably one of the dynamics that has saved China from inundation by squatter and shanty-town settlements.

A fifth observation, is that China's pattern of urbanization is unusually bi-polar and may well continue to be so in the future, with urban growth at both ends of the scale spectrum. Furthermore, this bi-polarity may present China with a different way out of social and environmental problems by allowing a virtuous variety of services and habitats to be provided at a variety of scales and without such a dependence on large cities.

A sixth observation is that urbanization and industrialization has been running well ahead of environmental carrying capacity, as economic production has superseded environmental remediation. Recent estimates of "environmental elasticity" (EE), for instance, which relates environmental change with economic change, shows China at –0.43 or weakly sustainable where 1% of economic growth is equivalent to 0.43% diminution in environmental quality, as compared to Netherlands = +0.04; Japan = –0.01; U.S. = –0.13. Also a recent cost of remediation estimate by China's Academy of Sciences knocks off about 4% from 9%～10% GDP growth numbers.

A seventh and final observation is that living density in China is high but, overall, declining in urban areas. Overall, the average population per square kilometer of 657 cities is 9,000 people per square kilometer, very close to the policy target of 10,000 people per square kilometer. This has given rise to internal dynamics within cities, including de-concentrating overcrowded and dilapidated areas, displacements often to urban peripheries for some, higher-rise building within cities, and side-by-side developments, eg. Pudong to Puxi in Shanghai. Indeed, in a comparative perspective, urban populations of top metropolitan areas for U.S., Italy, and China, shows a certain level of convergence, with densities in Italy and the U.S. moving up slightly although still well below those of China (Fig. 8).

Now, turning attention to estimates of forward projection. First, there is population growth, which seems likely to rise from around 1.31 billion to about 1.5 billion in 2035 before declining.

	1985-1990	1990-1995	1995-2000	2000-2005	2005-2010	Average
Beijing	2.5	1.4	3.3	3.3	2.6	2.6
Shanghai	1.8	2.2	2.4	3.8	1.3	2.4
Tianjin	1.7	0.6	1.1	1.3	1.2	1.2
Chongqing	1.7	4.0	6.7	4.6	3.4	4.1
Wuhan	2.1	2.7	3.3	2.7	2.7	2.7
Suzhou	2.7	2.0	2.7	9.7	2.5	3.9
Wenzhou	1.5	2.8	3.1	3.3	2.7	2.7
City Average	3.1	3.2	2.7	5.6	2.8	3.4

Seoul: 1960-1970 = 10%±; 1970-1980 = 6%±
Tokyo: 1900-1910 = 8%±; 1950-1960 = 5%±
U.S. City Ave. (1950-2005) = 1.8%±

图7 中国和其他地区城市增长率
（资料来源：中国统计年鉴，1985—2010年；美国人口统计局；东京都政府；世界人口综述）

Fig.7 Growth rates of Chinese and other cities
(Source: China Statistical Yearbooks, 1985-2010; U.S. Census Bureau; Tokyo Metropolitan Government, World Population Review)

每平方千米居住为9000人，与每平方千米10000人的政策目标非常接近。由此引发了城市内部的人口迁移，包括人口从过度拥挤和衰败地区向城市边缘地带疏解，或向高层住宅区转移，抑或是这两种情况的并行发展，例如上海浦东和浦西的发展。事实上，比较来看，美国、意大利和中国几个顶级大都市地区都呈现出一定程度的城市人口聚集现象，意大利和美国的城市人口密度略有增长，但仍远低于中国的水平（图8）。

现在，让我们将注意力转向对未来中国城市化发展趋势的预测。首先是人口的增长，有可能从13.1亿增长到2035年的15亿，然后开始下降。在1750—1850年间中国人口增加了一倍，1850—1950年间增长了23%，到1950—2050年间人口将再翻两番。这一预测值也被联合国的相关研究所证实。然而，这并不完全是个好消息：到2030年，年轻劳动力的数量会下降，老年人口将在20年间从1.65亿增长到2.4亿，适婚青年的男女比将是120∶100。

谈及未来城市化的多种方案，城市化率在21世纪中叶之前将达到65%～70%。这意味着中国将要产生200多个城市和至少1亿的城市居民，与拥有约250个城市、处于城市化发展成熟期的美国相比，某种程度上仍属于"欠城市化"，其城市化进程也并非前所未有（图9）。预计上海、北京等大城市的人口规模仍将继续增长，分别达到2700万和2300万。

对中国人口的预测研究显示，若按每平方千米10000人的标准，2005—2045年间增加的城市地区相当于110个浦东。从发展过程看，早期阶段扩张速度是最高的，之后城市化速度趋于平缓。到今天，中心地区变得更加饱

There was between 1750–1850 a doubling of population compared to an 1850–1950 increase of 23%, and then a 1950–2050 doubling again. This future estimate is also borne out by the U.N.'s study, however, with not entirely positive news: there will be fewer younger workers in the 2030 workforce; older population increases from 165 million to 240 million people in 20 years; and a marriage squeeze of 120 boys to 100 girls.

Moving through various scenarios of future urbanization, the overall proportion of urbanization seems likely to reach to around 65% to 70% by mid-century. This will yield some 200+ cities with at least 1 million inhabitants, compared to the mature U.S. equivalence of around 250 cities-still somewhat "under-urbanized" and nominally not unprecedented (Fig. 9). Also, large cities like Shanghai and Beijing can be expected to grow in population to somewhere near 27 million for Shanghai and 23 million for Beijing.

Plots of these population projections onto the map of China show an additional urbanization coverage equivalent to 110 Pudongs from 2005 to 2045, assuming a standard of 10,000 people per square kilometer. The rate of expansion is highest during the earlier periods, before tapering-off, as the rate of urbanization becomes asymptotic. Today's centers become more saturated, plus there will be corridor-like development in the center linking Beijing, Wuhan, and Guangzhou. Regarding the footprint of urbanization, overall, it represents roughly a doubling between 2005–2045. Needless to say, this urban expansion may be subject to a bundle of potential constraints. However, it seems likely that China will reach its mature level of urbanization in much less time than did the U.S. from a comparable starting position.

Now, looking below the national level and at one of China's more prosperous and developed regions, matters of urban transition and scale become more sharply defined and amplified. Here, we will look at the Changjiang Delta Region, extending some 350 km in from the coast and embracing Jiangsu, Zhejiang, and part of Anhui Provinces as well as Greater Shanghai (Fig. 10). This is a region that has a population of some 70 million of which 55%–60% are classified as urban dwellers, well ahead of the national statistics. In 2003, the region accounted for 18.5% of national GDP, 22% of

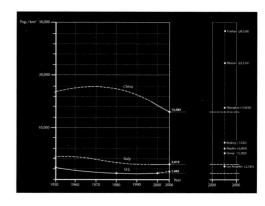

图8 中国、美国和意大利主要城市平均人口密度比较
（资料来源：中国统计年鉴；美国人口统计局；意大利统计局）

Fig.8 Comparison of average urban population densities of major cities in China, U.S. and Italy
(Source: China Statistical Yearbooks, various; U.S. Census Bureau; ISTAT)

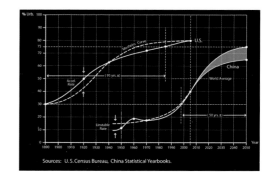

图9 美国和中国长期城市化率的比较
（资料来源：Peter G. Rowe）

Fig.9 Comparison of longer-term rates of urbanization between U.S. and China
(Source: Peter G. Rowe)

和，并且在北京、武汉、广州之间形成了廊道状的发展地区。总体而言，中国的城市化率从2005—2045年将大致翻一番。尽管这一城市扩张历程可能会受到各种潜在限制的束缚，但与美国相比，中国从类似的起始位置步入城市化成熟期所需要的时间要短得多。

现在，让我们来聚焦中国内部那些较为繁荣和发达的地区，城市转型和规模的问题在这里变得更加尖锐和放大。以长江三角洲地区为例，它覆盖海岸线向内大约350千米的范围，包括江苏省、浙江省和安徽省的一部分，以及大上海地区（图10）。这一地区的人口约有7000万，其中55%～60%为城市居民，远超出当时的国家平均水平。2003年，该地区贡献了全国GDP的18.5%，国家税收的22%，国家出口总额的28.4%，可以说是一个巨大的经济引擎。

从历史上看，长江三角洲地区因其优越的自然环境条件（如肥沃的土壤、充足的水源、温和的气候等），长久以来一直是良好的人类定居地。地区交通发达，在17和18世纪就拥有京杭大运河等交通干线，进入19世纪更是进一步蓬勃发展。一张1912年的地图很好地说明了地区主要中心、贸易集镇以及它们之间一部分的交通联系。1842年之后，随着通商口岸的出现，上海逐步成为一个主要的集散中心，沟通海上贸易与腹地的联系。至19世纪末，它成为了主要的工业中心，1900年前后居民人口100万，到1920年成为一个庞大的城市。其他繁荣的地区还包括如南京，在1928—1937年间曾作为首都。但是，也有一些传统城镇走向没落。例如太湖地区的南浔，人口从1900年的40000下降至1949年的13000。

不过总体而言，除少数个例以外，从20世纪早期到

national tax revenue, and 28.4% of the nation's export revenue ie., it is a big economic engine.

Historically, the Changjiang Delta Region has long been well-settled due to favorable environmental factors (eg., fertile soils, water, a moderate climate, and so on). Transportation has also been good, among other routes including via the Grand Canal. Certainly, by the 17th and 18th Centuries, let alone into the 19th Century, the region was thriving. A 1912 map gives a good sense of major centers, market towns and parts of the transportation links among them. With the arrival of the Treaty Ports, from 1842 onwards, Shanghai gradually became a major distribution center connecting sea-trade and the hinterland. Then, by the end of the 19th Century, it also became a major industrial center, although not becoming a very large city until around 1920, after reaching 1 million inhabitants by around 1900. Other places also flourished, Nanjing, for instance, was the capital between 1928 and 1937. However, a number of traditional towns also fell into decline. Nanxu, for instance, in the Lake Tai area, dropped in population from 40,000 in 1900 to only 13,000 people in 1949.

By and large, however, the spatial distribution of cities and towns did not change appreciably, with few exceptions, from the beginning of the 20th Century to the early 1950s, not a great difference between a map of 1951 and the one for 1912. The post-1978 socio-economic reforms, however, led in another direction: (1) surplus labor, released by the modernization of agriculture, became absorbed into Township Village Enterprises; (2) higher proportions of local revenues were retained locally; and (3) local authorities obtained more control and responsibility over their own economic development. Then came the 1989 amendment of *Land Administration Law*, and the first experiments in Shenzhen of leasing "use-rights" as a local revenue stream, followed shortly thereafter by a national program.

Transition in the Changjiang Delta to the current pattern of urbanization, however, did not occur all at once, nor in the same direction, as illustrated by data on migrational patterns. Between 1964 and 1989, for instance, the whole region was a magnet. From 1989 to 1993, the dominant destination lay at the center in the Lake Tai area. From 1993 to 1998, migration

图10 长江三角洲地区的城镇分布
（资料来源：Peter G. Rowe）

Fig.10 Distribution of cities and towns in the Changjiang Delta Region
(Source: Peter G. Rowe)

1950年代初，长江三角洲地区城镇的空间分布并未发生显著变化。将1951年与1912年的地图相对比，也可以看出差异不大。而1978年以后的社会经济改革则带来了翻天覆地的变化：（1）农业现代化解放出大量的农村剩余劳动力，被村镇企业所吸收；（2）地方财政收入保留在本地的比例增加；（3）地方政府对当地经济发展的控制和责任增强。1989年《土地管理法》修正案颁布实施，深圳创先通过租赁"土地使用权"作为地方财政收入的来源，随之推广至全国。

关于迁移的统计数据显示，长江三角洲地区的城市化模式，并非一夜之间形成，也不是一成不变。举例来说，1964—1989年间，整个区域如同一个巨大的引力中心；1989—1993年间，太湖地区成为最主要的迁移目的地；1993—1998年间，人口向大城市聚集；1998年至今，大城市的集聚效应和小城市的扩张同步出现。

在最近一段时期，这个区域的城市领土确实有所扩张，但从2003年的人口分布图可以看出，同时出现了一系列密集分布的乡镇聚居点。简言之，长江三角洲地区经历了一个分阶段的城市化进程：（1）初期阶段，产业和城市定居点蓬勃发展；（2）之后人口向地区的中心城市迁移，再后转向较大的城市地区；（3）大城市人口向边缘地带疏解，同时伴随区域内一些小城镇的发展。

这一演变过程中至少呈现出四种典型的城市形态和动力：（1）在地区北部和西部的独立的城市扩张；（2）大型区域扩张的两阶段过程，先是人口聚集，然后向外扩张伴随以卫星城发展；（3）主要沿沪宁高速公路形成城市走廊；（4）城市区域的合并。

长江三角洲地区1950年和2010年的公路网情况对比

was concentrated towards large cities. Then, from 1998 to the present there was a continuation of the concentration plus expansion of smaller cities.

Over the recent period, cities have certainly expanded their presence, but there has also emerged a dense array of smaller settlements in the form of towns and townships in the region, as shown in the scale population distribution from 2003. In short, what appears to have transpired in the Changjiang Delta was a staged process of urban formation with: (1) a nascent stage of booming in-sector development and urban settlement; (2) then migration of population first to the center of the region and then into the larger urban areas, followed by; (3) decentralization of very large cities into peripheries, alongside of growth of some smaller towns within the region.

Also apparent in this transitional process were at least four broad urban forms or dynamics: (1) independent urban expansion, in the north and west of the region; (2) a 2-stage process of expansion of very large areas, including the attraction of population followed by outward expansion and generation of satellite developments. Then, there was (3) the formation of urban corridors, principally along the Shanghai-Nanjing expressway, and (4) consolidation of urban areas.

Indeed, infrastructure developments, that have been considerable, had much to do with these trajectories, comparing the 1950 and 2010 roadway networks. The result has been a significant upgrading of transportation accessibility between 1950 and today and, therefore, of development potential. Note that between 1950 and well into the 1980s there was no real way across the Changjiang, hence the isolation of the northern side. For 2010, note that Anhui Province is still relatively isolated in this time-distance depiction (Fig. 11).

One outcome has been the wholesale conversion of agricultural land into urban land, resulting, unfortunately, often in over-development and redundancy with non-complementary activities among smaller communities. Also, there has been a loss of arable land. It should also be noted that transportation access is but one measure of distancing of one place from another. Other forms of measurement include cultural distance, administrative distance, physical distance, and economic distance. At the "small" end of the urban scale, it is also

可见，基础设施建设对这些发展轨迹的影响已相当显著。与1950年相比，今天这个区域的交通可达性得到了显著提升，因而发展潜力也有了很大升级。需要注意的是，1950—1980年代期间，一直没有真正的跨江通道，江北地区因而被孤立。直到2010年，由于时间-距离的关系，安徽省仍然处于相对孤立状态（图11）。

城市扩张的结果之一是大规模的农用地转变为城市用地。但这往往导致了过度开发，以及同质化小型社区的重复建设，此外还伴随以耕地流失的现象。需要注意的是，交通可达性只不过是度量两地之间距离远近的方法之一。此外还有很多其他的测量方式，包括文化距离、管理距离、空间距离、经济距离等。对于那些处于城市规模等级末端，人口数目少于50000的小型城市中，经济、就业和服务能力呈现出显著差异，例如GDP水平以及供水、排污等城市服务。考虑到50000人以下的乡镇的数量、分布以及可能合并的需求，这个问题绝非无关紧要。诚然，上海周边的一些城镇，诸如同里镇和梅李镇等，因旅游业的繁荣和作为第二居所而发展迅速。但还有大量的小城镇因为难以实现基本的规模效益，而无法做到全面自给自足的发展模式。而事实上，恰恰是这些城市与大型乡镇之间的城市化区域使得中国城市化的巨大规模得以真正显见。

为了能够直观感受这个规模之巨大，我们利用遥感数据图像进行时间序列下城市发展的分析。很显然，1950—1975年间，城市的发展变化不大；而1950—2010年间则呈现出翻天覆地的变化，出现了真正称得上是巨型区域化的城市发展（图12）。此外，例如在常州，城市开发和功能布局通常呈现出无序的蔓延。仔细观察，可以发现这不同于我们通常在美国城市或者东京所描述的

apparent that economic, employment, and service capacities display disparities below 50,000 people, e.g. GDP, and urban services like water, sewage. This is not a trivial issue, when considering the number and distribution of townships of less than 50,000 people, and the need, potentially, for consolidation.

Admittedly, some towns e.g., Tongli and Meili, within driving range of Shanghai have prospered through tourism and as second home markets. Nevertheless, the incapacity for many small towns to achieve basic scale efficiencies has called into question the across-the-board model of self-sufficient. Indeed, it is precisely in the conurbated, urbanized areas between cities and larger towns that the vastness of China's urban scale becomes truly apparent.

To give a flavor of this vastness, images of a time sequence of urban development derived from remote-sensed data were analyzed between 1950 and 1975, and it is apparent that it is not very great in the Maoist period. However, looking at it from 1950 to 2010, we quickly see this massive vastness, the emergence of what probably really qualifies as mega-regional urban development(Fig. 12). Moreover, as in the case around Changzhou, the urban pattern of use and development is often sprawling and indiscriminate. Looking closer, the spread is, in fact, different from say "sprawl" in U.S. cities or even in Tokyo. It is (1) denser; (2) largely comprised of "industrial parks", which are often unforgiving of local circumstances, frequently duplicative, functionally inefficient, and environmentally disastrous. Part of the problem is also jurisdictional.

Analytically, this dispersion and fragmentation of urban development can be subject to spatial analysis of environmental suitability using a basic Boolean method of screening among attributes or characteristics of the environment, along with intersection with urban development. The resulting gradient map shown here, where green = suitable, and red = unsuitable, is overlaid with a map of urban development – the black dots (Fig. 13). It shows a reasonable result, although unsuitably in many developed areas, but with the most unsuitable development internalized by particular jurisdictions. However, they are often (1) focused elsewhere; (2) too weak in capacities; and (3) confronting incursions by neighbors straddling developments.

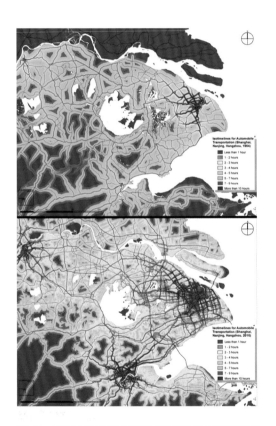

图11 长江三角洲地区1950年与2010年的交通可达性比较
（资料来源：Peter G. Rowe）

Fig.11 Differences in transportation accessibility with the Changjiang Delta between 1950 and 2010
(Source: Peter G. Rowe)

"蔓延"，它体现为：（1）更加密集；（2）主要由"工业园区"构成，并且往往不尊重当地条件，重复性建设、效率低下且环境污染严重。这些问题部分是由行政原因造成。为了剖析这种分散的、碎片化的城市发展，我们利用布尔运算筛选环境属性和特征，进行环境适宜性的空间分析，然后计算与城市开发用地的交集，由此得到梯度图，其中绿色代表适宜，红色代表不适宜，与城市开发用地重合的部分就是这些黑点（图13）。分析显示，在发达地区通常不适合开发的地区包括：（1）非重点开发区；（2）承载力较弱地区；（3）相邻地区已有强劲发展。

一个有趣的发现是，社区越小，或者更确切地说，由越小定居点形成的集聚区，在农田、森林、湿地、淡水水体等资源方面的损害情况越糟糕。为此我们进行了一项研究，选取在人口密度超过1000人／平方千米的230个镇范围内，测度各类资源损失与不同规模定居点之间的关联性。

进一步缩小到城市尺度来看，在不同地区，比如历史地段、高密度的繁华商业区、特殊企业发展区，或是特别经济区等，城市化的建设模式和空间形态呈现出高度的差异化特征。类似的情况也延伸到内城邻里、郊区地带和卫星城、城郊周边的城市化地区、城市中心区内部及其周围的城中村。可以说，中国自1949年以来的城市化进程是通过一连串的造城运动和项目实现的。这里指的是中观尺度的发展，介于宏观尺度的总体规划布局和微观尺度的特定建筑物、建筑群之间。它通常涵盖大量的住宅建筑以及其他相关功能和商业服务，是当前的常见情形，比如在上海的浦东和贵阳的金阳。

上述现象并不算新奇，类似情况也在世界其他地区的

In all of this it is also interesting to note that it is the smaller communities, or, more precisely, the aggregation of the smaller settlements which perform worst with respect to the loss of resources like agriculture, forests, marshes, freshwater bodies etc. An analysis was conducted that began by identifying losses of various resources in relationship to settlements of various sizes. This was followed by the loss of resources within the radius of some 230 towns with distinctions of more than 1,000 people per square kilometer.

Moving still further down in urban scale, the constructed manifestation and physical form of Chinese urbanization is by no means uniform, ranging across a variety of environments, including historic areas, dense and bustling commercial areas, special enterprise developments, and special economic zones. Manifestations also extend to inner-city neighborhoods, peripheral and satellite developments, urbanization in peri-peripheral areas, urban villages both in and around urban centers. However, it is also fair to say that the bulk of China's urbanization since 1949, if not before, has occurred through a succession of urban district-making activities and projects. Here, I am referring to meso-scale developments, between the macro-scale of overall plan layouts and the micro-scale of specific buildings and building complexes. Typically involved are substantial residential components alongside other related uses and commercial services as can be seen in contemporary circumstances like Pudong in Shanghai and Jinyang in Guiyang.

This phenomenon is by no means novel but analogous to other practices in other parts of the world at various times, including Cerda's Eixample or extension to Barcelona from the 19th century onwards; the quartiere of Rome, mainly after the 1870 Risorgimento; the residential estates of Singapore, of post-colonial vintage; and the grid-iron of American suburbs, as exemplified in post-World War II Los Angeles. In China, this phenomenon has been and still is largely determined by overlapping procedural planning processes, regulations, control detail plans, and other forms of prescriptive technical specification. The upshot is a consistent "image" or "template" controlling the production of urban space. Moreover, often what is specified is what you get, as in the case of Zhengzhou

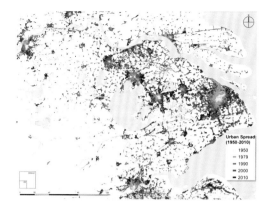

图 12　1950—2010 年间长江三角洲地区的城市扩张
（资料来源：Peter G. Rowe）

Fig.12　Urban expansion in the Changjiang Delta between 1950—2010
(Source: Peter G. Rowe)

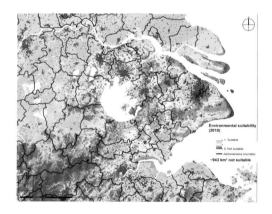

图 13　长江三角洲地区城市化进程中的相关土地覆盖情况和环境适宜性评价
（资料来源：Peter G. Rowe）

Fig.13　Relative land-cover and environmental suitability of urbanization in the Changjiang Delta Region
(Source: Peter G. Rowe)

不同时期发生着，包括19世纪塞尔达在意大利巴塞罗那实施的扩建，1870年复兴运动之后罗马旧城内的街区建设，新加坡后殖民时期的房地产发展，以及"二战"后洛杉矶为代表的美国郊区的格网式蔓延。在中国，这种现象曾经并仍然在很大程度上取决于相互叠加的一系列程序化内容，包括规划过程、制度法规、控制性详细规划以及其他形式的专业技术规范。结果是某种高度一致的"图景"或"模板"控制了城市的空间生产。而且，往往会对最终的空间成果进行明确限定，郑州就是一个典型案例（图14）。

更为普遍的是，城市设计在中国体现出的基本方法似乎一直固守"现代主义"城市-建筑的特征，具体体现为：（1）大型街区；（2）单元式的开发模式；（3）低质量的道路和其他基础设施建设；（4）高度同质化的建筑类型；（5）强制性的高度的自给自足。对比历史，这种空间组织方式可以视为传统街区建设方法的延续。例如清代北京城随处可见宽阔的大道、飞地式的封闭社区以及十分有限的建筑类型。然而，尤其在当前，面对层出不穷的新的城市问题，这种城市建设模式遭遇到了诸多挑战。

第一，占主导地位的低质而宽阔的道路网络，相比于其他地方的街区建设，显然更容易导致交通拥堵和隔离，以及相关成本的提高。另一方面，在某些情况下，可以对宽阔的道路系统善加利用，例如昆明建设BRT系统，或是布局基础设施改造工程（图15）。

不幸的是，地毯式铺开的建设模式可能带来对生态环境（例如地表覆盖、水资源保护等）和历史景点的不利影响。例如在武汉最近就出现了类似的状况。另一方面，超

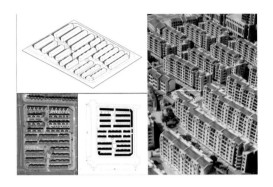

图14 中国式超大街区"模板"
（资料来源：Peter G. Rowe）

Fig.14 Chinese "template" of superblock development
(Source: Peter G. Rowe)

图15 交通拥堵和部署BRT设施的机会
（资料来源：Peter G. Rowe）

Fig.15 Traffic congestion and BRT installation opportunities
(Source: Peter G. Rowe)

(Fig. 14).

More generally, the underlying approach in urban design seems to exhibit sustained adherence to "modernist" urban-architectural traits, eg. (1) large block configurations; (2) a cellular pattern of development; (3) coarse-grained deployment of roads and other infrastructure; (4) considerable uniformity in building types; and (5) a high degree of imposed self-containment. Looking back in time, aspects of this spatial configuration also adhered to traditional manners of district-making. Qing Beijing was no stranger to wide thoroughfares, nor was it a stranger to enclave-like gated communities, nor to a relatively limited typological repertoire of building.

At this juncture in time, however, a number of rather pointed questions can be raised about the fittedness of the template in play, especially for contemporary and emerging urban circumstances.

First, the predominance of a coarse-grained and wide roadway network, with not much in between comparative to other cases of district making, clearly lends to traffic congestion and isolation, along with associated costs. On the other hand, in some cases, broad rights-of-way can be put to good use, like the BRT system in Kunming, as well as locations for other infrastructural retrofits (Fig. 15).

Unfortunately, carpet-like deployment of the 'template' can run afoul of environmental considerations regarding land cover, water conservation, as well as historic and scenic sites. Until recently, this was the case in Wuhan. On the other hand, the leeway provided by the sheer expanse of these superblock estates provides room, in some cases, for water harvesting and rainfall runoff control. The 'superblock template' is also awkwardly positioned with regard to dynamic processes of local-level despecialization that flow over cities, allowing them to restructure and become better places.

Another way of looking at a similar aspect of urban formation is through the lens of diversity alongside of compactness at the meso- and macro-scale. Further, at first glance, both characteristics would seem to represent a virtuous combination with regard to diminished environmental impact. However, this may not be the case. For instance, high levels of diversity but uneven distribution can enlarge travel distances. Also,

大街区蔓延的同时也留下不少空余之地，在某些情况下，可以用于水资源回收和降雨径流控制。这些"超大街区模式"由于过度强调功能分区，难以适应城市的动态发展，所以需要通过重新调整使其成为更积极的场所（图16）。

另一种观察城市形成方式的类似方法则是在中观和宏观尺度下考察城市的多样性和紧凑度。乍看之下，这两个特征在降低环境影响方面似乎代表了一种良好的组合。然而，事实也许并非如此。例如拥有高度的多样性但分布不均，可能进一步扩大交通距离。此外，根据城市经济学的"马歇尔困境"，除了为提高经济绩效而必需的集聚经济以外，多样性也同样具有优势。

在对武汉近期发展的研究中，通过整理各城区土地利用多样性的数据，用跨区交通的估算值作为城市紧凑度的表征，然后对两组数据进行相关性分析，发现它们在低值和高值区间端都呈现出非线性的特征。问题似乎来自在城市边缘区和中央商务区都采取了相对同质化的发展模式。此外，还包括行政管理权的碎片化，以及指导政策和协调激励机制的缺乏。许多的城市住宅区，特别是临近郊区的地方，服务功能匮乏，并且超出了目标人群的可支付能力（图17）。

针对上述问题，这里提出两大改善策略。第一，基于理想的、适宜步行的建筑尺度，规划布局140米×140米的小规模街区。事实上，已有大量的实践致力于让中国告别对超大街区的依赖，以及由此带来的诸多浪费，包括彼得·卡尔索普在昆明呈贡新区，詹姆斯·科纳事务所在深圳前海地区，吉姆·布里尔利及其BAU事务所的网络城市，以及维托里奥·格里高利在上海浦江新城的规划实践。

according to one of Marshall's dilemmas in urban economics, diversity is advantageous except when aggregations of economic scale are required for better economic performance (Fig. 16).

In the case of Wuhan and its recent development, data were compiled on land-use diversity by district; inter-district travel data was estimated as a proxy for compactness; relating the two there was a tendency at both the low and high ends to exhibit non-linearity. And, indeed, the problems seem to come from relatively homogeneous developments on the urban periphery and in the central business district. Another related issue comes about from fragmentation of authority plus a lack of well-directed and coordinated incentives. Many urban residential districts, especially towards the periphery, are under-serviced in addition to being out of affordable range of target populations (Fig. 17).

Now, as far as redress is concerned, two broad strategies appear to offer better outcomes. The first is to plan with smaller block and grid sizes, based ideally on optimal blends of viable building footprints, nominally of 140 meters by 140 meters. Indeed, a number of efforts to wean China away from an over-dependence and wasteful use of superblocks have occurred, including Calthorpe in Chenggong New Town in Kunming; James Corner Field Operations at work in Qianhai, Shenzhen; James Brearley and BAU's Networks City approach; and Vittorio Gregotti at work in la Nuova Citta di Pujiang shown here.

A second strategy involves deployment of urban-architectural spatial techniques within existing large-block enclaves. There are at least five kinds: (1) hybrid building ensembles eg. the Linked Hybrid by Steven Holl in Beijing; (2) field operations eg. Shinonome Canal Court by Riken Yamamoto, Tokyo; (3) super-sized buildings, such as "The Whale" by de Architekten Cie, Amsterdam; (4) "lines" and "rails", such as the Il-Plein scheme by OMA in Amsterdam; and (5) use of "webs" and "mats", with few examples except for Tuscolano by Libera (Fig. 18). In short, though, the time-honored template for urban district-making in China needs to be reconsidered and revised emphasizing flexibility, expanded functionality, more redundancy, and potential for inter-temporal change and adaption.

Finally, if there is a punchline to these stories, it is that China

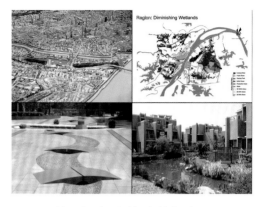

图16 环境影响、水源保护和场地发展机遇
（资料来源：Peter G. Rowe）

Fig.16 Environmental impact, water conservation and on-site opportunities
(Source: Peter G. Rowe)

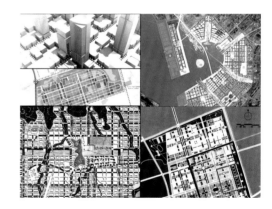

图17 各种模式化的超大街区开发
（资料来源：James Brealey and Fang Dan, *Networks Cities*, 2011）

Fig.17 Alternatives to "template-like" superblock development
(Source: James Brealey and Fang Dan, *Networks Cities*, 2011)

第二，建议将城市-建筑的空间设计手法引入现有的大型街区中，至少有以下5种方式：（1）混合式建筑群，例如史蒂芬·霍尔在北京的当代MOMA；（2）场地设计，例如山本理显在东京的东云运河苑；（3）超级建筑，例如de Architekten Cie事务所在阿姆斯特丹的"鲸鱼"；（4）"线"和"轨"，例如OMA在阿姆斯特丹的IJ-Plein广场设计；（5）使用"网"与"垫"，例如利伯拉的Tuscolano项目（图18）。简言之，需要对中国城市街区建设中长久以来形成的固有模式进行重新设计和改进，强调灵活性和功能拓展，从而更好地应对未来变化。

若为上述问题做个总结，那就是中国正处在多维度城市演进的交叉路口。首先，总体而言，传统的能源和资源高耗型的发展道路是不可持续的。事实上，中国似乎应该寻求介于过度开发和欠开发之间的"第三条道路"。其次，在区域层面，最大的问题出在较大规模的建制镇之间及其周边地区。不过，若能利用城市规模分布两极分化的优势，将可能探索出一条灵活的、特色化的、高度网络化的发展道路。最后，在更加微观的层面，中国需要跳出超级街区的建设模式，转向更灵活和可持续的新路径，不会因很快过时而造成资源浪费。

appears to be at something of a cross roads with regard to urban formation and at a variety of levels. First, at an overall level, conventional energy and resource intensive pathways forward are probably not sustainable. Indeed, it seems as if China should be seeking a "third way" between over- and under-development. Second, further down in the regions, the big problem lies in between and around bigger designated towns and regions. However, by taking advantage of bi-polar formation at either end of the scale spectrum, flexibly specialized and highly networked arrangements might be imagined, pointing to an alternative pathway forward. Third, as mentioned, at a lower level still, China needs to move away from its basic superblock template of development into something more flexible, able to be readily despecialized and likely to be more sustainable, with far less costs due to obsolescence.

图18 应对大型街区的多种策略
（资料来源：右上 Peter G. Rowe；其他 Peter G. Rowe and Har Ye Kan）

Fig.18 Alternative schemes for large-block enclaves
(Source: top right by Peter G. Rowe; others by Peter G. Rowe and Har Ye Kan)

参考文献 / References

BREALEY J, FANG D, 2011. Networks Cities.

LU J H, ROWE P, ZHANG J, 2001. Modern Urban Housing in China: 1840-2000[M]. Munich: Prestel.

ROWE P, 2006. East-Asia Modern: Shaping the Contemporary City[M]. London: Reaktion.

ROWE P, 2011. Urban Residential District Making[M]//The peoples Republic at Sixty: An International Assessment[M]. Cambridge, Mass.: Harvard University Press: 163-172.

ROWE P, FORSYTH A, KAN H Y, 2016. China's Urban Communities: Concepts, Contexts and Well-being[M]. Basel: Birkhaüser.

ROWE P, GUAN C H, 2016. Striking Balances Between China's Urban Communities: Blocks and their Layouts[J]. Time + Architecture, 6: 29-34.

ROWE P, 2017, Spatial Practices in China's Urban Residential Districts: A Time for Change[M]//Letters to the Mayors of China. New York: Terreform: 206-221.

ROWE P, KAN H Y, 2014. Urban Intensities: Contemporary Housing Types and Territories.

东亚城市的新兴建筑地域

EMERGENT ARCHITECTURAL TERRITORIES IN EAST ASIAN CITIES

彼得·罗
哈佛大学杰出贡献教授
哈佛大学设计学院教授
清华大学建筑学院荣誉教授

Peter G. Rowe
Distinguished Service Professor, Harvard University
Raymond Garbe Professor of Architecture and Urban Design, Graduate School of Design, Harvard University
Honorary Professor, School of Architecture, Tsinghua University

中文翻译：刘佳燕，张琪
Chinese translation by Jiayan Liu and Qi Zhang

聚焦当代的东亚城市,特别是其近15年以来的发展是极具启发性的。第二次世界大战之后,东亚地区发展迅速。早期阶段,随着日本和英国的殖民统治及霸权主义的崩塌,大规模现代化成为独特的驱动力。到处弥漫着一种由上而下、专制的和寡头的氛围。一切以狭义的生产为导向,强调社会控制,以期实现社会稳定和物质改善(图1)。

在物质空间层面,这种现代化采取了国际普遍的发展模式,主要效法西方的做法,中国则主要受苏联的影响。具体体现为总体规划的实践、使用和生产资料的理性集中分配、面向基础设施的大规模投资、集体消费的推行、粗放低质且缺乏弹性的规制,等等。总体而言,这些策略大体是成功的。在多数地方GDP持续增长,主要城市呈现出现代化、高密度、规模化的风貌,并拥有了更多更好的服务。而且时至今日,上述状态仍在很多地方延续。不过,自1990年代以来,一系列事件使得当权者以及其他社会阶层放弃了原来狭隘的现代化道路,转向接纳更为广泛而且丰富的生活方式,提升环境友好度,以及设立更高的城市物质生活标准(图2)。新加坡就是一个典型案例。随着社会整体富裕程度的提升,以及人们的愿景和品味的提高,这些趋势进一步加剧,同时也是响应城市更具竞争力的发展需求,以吸引自由投资者和专业化劳动力。

另一方面,这片地域上同时呈现出一种城市动态演变过程的范式转换,也可以说是在一段长期而流畅展开的历史长卷中的"转折点"。类似情况曾经被多位历史学家所描述,例如米里亚姆·莱文和她的同事关于第二次产业革命的论述,以及尼尔·弗格森最近对于美国面临迅速崩溃的推测。相关观点也可见于其他学科的讨论中,例如经济学家借助空间维度的分叉现象研究"去地域化"和"再地域化"

图1 以生产为导向的发展
(资料来源:Peter G. Rowe)

Fig.1 Production-oriented developmental processes at work
(Source: Peter G. Rowe)

图2 香港的城市转型
(资料来源:Peter G. Rowe)

Fig.2 Urban transition in Hong Kong
(Source: Peter G. Rowe)

In looking at East Asian cities in contemporary times, and certainly over the past 15 years or so, it is instructive to begin seeing them in relation to developments more immediately in the aftermath of World War II, to which they are almost direct reactions. This earlier period was involved with the breakdown of colonial occupations and hegemonies – moving mainly away from Japan and Britain – and a time of a particular kind of drive towards wholesale modernization. At work, throughout, was a conspicuous form of guidance and one that was largely top-down, autocratic or oligarchic. It was narrowly production-oriented in focus and with a strong emphasis on allowance for social control in return for stability and material betterment (Fig. 1).

Physically, it was materialized via internationally-available development procedures, largely of western origin, including Soviet influence in the case of China. At work were master planning exercises; rational centralized allocations of uses and productive resources; heavy investment, when affordable, on infrastructure; an embrace of collective consumption; and deployment of broad-brushed, inelastic coarse-grained regulation. Overall, the general recipe was successful. GDP was on the rise in most places and major cities became modern looking, dense, relatively large and reasonably serviced. Indeed, moving forward in time, this state of affairs continued in many places. However, beginning in the 1990s, various concatenations of events led those in power, along with many other segments of society, to turn away from the narrow pathways forward in modernization towards a much fuller inclusion of broader lifestyle opportunities, improved environmental amenity, and higher material standards of urban living (Fig. 2). Certainly this was the case with Singapore. These trends were egged on by rising affluence, plus a broadening of aspirations and tastes, as well as in response to the need for cities to become more competitive with regard to footloose investors and, of course, participants in evolving and more sophisticated labor forces.

Moreover, what appears to have transpired in the region is something of a paradigm shift or a "turning point" in the dynamic of urban change- an interruption or inflexion point in the longer term and smoother unfolding of history. This kind of phenomenon has also been described by a variety of historians like, for instance, Miriam Levin, and her colleagues with regard

共同作用的结果，又如有学者基于福柯的理论，借助文化论述中的反应和转变等概念，重新定义了什么是可接受的语言，以及当代规划领域中值得和不值得探讨的问题。

1980年代末期，韩国民主运动兴起，并在1993年达到高潮。1990年韩国还是一个相对贫困的国家，人均国内生产总值（购买力平价）约5000美元，如今已稳步增长到35000美元。首尔早期快速的城市现代化进程很大程度上受到军国主义的影响，从2002年开始转变为市长选举制，关注环境改造、历史保护，以及生活方式的表达。一个经典案例就是市中心的清溪川修复工程。在业已实现现代化和富裕的日本东京则是另一走向。1980年代形成的经济和房地产泡沫在1990年代初突然破裂，令东京和日本陷入困境，跨国竞争力急剧下降。在之后的2000—2004年间，日本为恢复领先地位做了大量努力，例如2002年内阁总理大臣办公室签署颁布《都市再生特别措置法》，并划定"都市再生紧急整备地域"。

新加坡和中国香港、台湾城市发展的转折也主要由内部事件驱动，同时伴随以由单一的"发展态"转向"竞争态"的全球性趋势。香港在1997年回归中国之后，通过重大基础设施改善等一系列工作推动城市品牌的重塑，最具代表性的就是位于赤鱲角的香港机场核心计划。中国台湾在1987年取消了《戒严令》，之后1990年代期间城市自治权不断得以强化，当局针对以往被忽视并逐渐激化的各类城市问题采取了系列举措。新加坡在经历经济衰退后，人民行动党作为执政党在1991年选举中遭遇挑战，这个海岛国家的家长式政府及时调整，转变曾经过于严格的姿态而走向开放，致力于营造宜人的、多样化的生活环境，全力迈向"热带卓越城市"和"活力宜居新加坡"的目标（图3）。

to the Second Industrial Revolution, or recently by Niall Ferguson, speculating about the potential of rapid demise in the United States. It can also be seen from the perspective of other disciplines in, for example, the outcomes of simultaneous processes of "deterritorialization" and subsequent "reterritorialization", via bifurcation phenomena in spatial dimensions familiar to economists. Or, via reactions and corresponding shifts in cultural discourse, in Foucault's sense, redefining acceptable language and, therefore, what is discussable and what is not in contemporary planning, and so one could go on.

As far as places and a sampling of events are concerned, in Seoul and South Korea, one can begin in the late 1980s with the final rise of the democracy movement, becoming more fully embraced by 1993. A relatively poor country as recently as 1990, at around $5,000 per capita, general affluence rose steadily to around $35,000 per capita recently in purchasing power parity terms. Seoul's rapid modernist urban development largely under militaristic guidance, gave way, certainly by 2002, to orderly elected mayoral transition and brought with it an emphasis on environmental amenity, conservation of various kinds, and lifestyle expression, symbolized as much as anywhere, by the Cheonggyecheon Restoration project in the heart of the city. In Japan and Tokyo, already modern and wealthy – events went in another direction. The economic and real-estate asset bubble that had been building in the 1980s suddenly burst in the early 1990s, plunging Tokyo and Japan into an obvious crisis, including a substantial loss of competitiveness almost across the board. Later, between 2000 and 2004, strenuous efforts were made to regain prior positions, including the enactment of Urgent Improvement Zone legislation in the realm of property development in 2002 emanating from the Prime Minister's Office.

Turning points in Singapore, China's Hong Kong and Taiwan were driven largely also by internal events, as well as by a general drift away from being solely 'development states' into becoming "competition states" within a far more global scheme of things. More specifically, in Hong Kong, there was the "hand back" from Britain to China in 1997, followed by a re-branding of the former city sate, involving a number of undertakings not least of which was Chek Lap Kok and the Airport Core Programs, including major infrastructure improvements. Taiwan saw release

图3 从发展态到竞争态的新加坡
（资料来源：Peter G. Rowe）

Fig.3 From developmental to competition state in Singapore
(Source: Peter G. Rowe)

可以肯定的是，中国的发展呈现出两面性：一面是成果辉煌的现代化发展，另一面则是广袤而相对落后的农村大地。但是最近，传统一刀切式的发展路径基本被摒弃，不再一味强调基本生产，而是因地制宜地探寻地方独特的比较优势。不过1990年代中后期至21世纪初的这段时期内，众多城市似乎仍然沿袭着相对传统的城市化模式——强调城市功能和建设规模的扩张，以及基础设施建设和投资。

如今，伴随上述城市发展建设的转型，特许开发的新区不断涌现，新的或者说有别以往的城市-建筑布局及表达形式也相继兴起。这里我用的"地域"一词有两重含义：其一是指类似北京CBD这样的一片区域；其二用来形容一种活动的领域和氛围。同样以北京CBD为例，强调的是高层建筑环绕中的居家和工作环境的营造（图4）。地理概念所反映出的物质要素的差异化，既可以是涵盖于某一地域范围内，又可以是包含在一系列活动内的——后者由实际或未来的建成环境，抑或内在的地理条件所决定。以高层建筑区为例，这种内在的地理条件可能由一系列的建筑形态和对其可能的解释所定义，如体型轮廓、环境绩效、自我调控性、材料统一性、计划协调性，等等。在实践中，可以通过建立一种情境逻辑的方法对任何特定的城市-建筑项目进行讨论和评价，具体包含以下"相互链接"的要素，如转折点、重要的规划设计方案、潜在的话语，以及包含双重含义的特定地域。

在当前这个从城市到建筑都多产的时代，如果只是意在勾勒以及举例说明在东亚城市化"后转折点"时期公共建筑的发展，我们似乎无法也没有必要系统地描述出所有的地域类型。由此，我将阐述六个小故事，对应于六种

from martial law in 1987, followed by greater municipal autonomy in the 1990s involving focused dealings with many of the deferred and mounting urban problems. In Singapore, following economic recession, there was a significant loss of votes by the People's Action Party in 1991 – the island state's paternalistic regime – leading in time, to less straight-laced attitudes, as well as conspicuous expansion of leisure-time opportunities, additional urban-amenities, and diversified living environments – all pushing strongly in the direction of a "City of Tropical Excellence" and to a "Lively and Livable Singapore" (Fig. 3).

To be sure, there are still at least two sides of China–one contemporary and glitzy and the other poor, largely agrarian and backward. More recently, however, there has been a substantial pulling away from "one-size-fits-all regimentation", accompanied by less overt emphasis on primary production; new searches from place to place for comparative advantage. Certainly since the mid- to late-1990s, and into the past decade, much urbanization of scale seems to have followed a relatively conventional pathway forward – one of expanded functionality, size of city building, and infrastructural heft and investment.

Now, within the scope of these turning points, new territories have been chartered and new, or different, forms of urban-architectural arrangement and expression have been engaged in. Moreover, territory here has two senses. First, reference to a tract of land or parcel like, for instance, the CBD in Beijing, as depicted here. And second, reference to a field or sphere of action. Again in the Beijing CBD, it is primarily concerned with high-rise building and the making of home-work environments (Fig. 4). Geography refers here to both differentiation of physical features within a territory, as well as within a course of action, where it is defined by what was actually built, as well as – potentially – by what could have been built, or its intrinsic geography. In the territory of high-rise building, for instance, this intrinsic geography might be defined by a range of possible construals and architectural accomplishment, like: shapeliness, environmental performance, self-regulation, material integrity, programmatic inclusion, and so on. In effect, for any particular urban-architectural project, a situational logic can be set up for discussion and assessment purposes by "chaining together", so to speak, turning points, the inevitable plans, and underlying

图4 北京 CBD
（资料来源：右上北京城市规划设计研究院，1993；其他 Peter G. Rowe）

Fig.4 From territories to projects in the Beijing CBD
(Source: top right by Beijing Municipal Institute of City Planning 1993; others by Peter G. Rowe)

不同的发展类型。首先，可以将其归纳为三种关于地域化的普遍理论，即：（1）向外延伸，并克服距离等障碍；（2）为新的发展创造离家更近的空间；（3）决定随着时间的推移保留和放弃什么，以及如何保留。

我希望先从克服距离这个话题开始。"飞入与飞出城市"，引发到达、离开以及连接等活动的展开，由此带来对机场及其设备等相关地域的关注（图5）。其背景是近年来东亚航空运输业的迅猛发展，这里已成为世界第二大空运市场，伴随大量新型设施的建设，年均旅客吞吐量高居世界首位。由此引出了我要讲的第一个故事——基于新建机场的区域创新和发展。其原型来自巴黎机场公司的建设，特点不仅体现在范围和规模上，而且还包括空间布局、朝向、流线调整、建设品质以及子系统整合等方面。

这类新机场的一个早期案例是位于日本大阪西南40千米建于人工岛屿上的关西国际机场，于1990年代初投入运营。项目方案征集采取国际竞赛的方式，最终伦佐·皮亚诺建筑工作室胜出。不过最终是由巴黎机场公司参照运输管理部门的要求，对原来的任务书进行调整，重新规划和确定了新的空间形态。其核心理念在于将国际和国内航班安排在一个多层多功能航站楼的不同楼层，取代过去将两者同层并置的组织方式，从而有利于实现包含换乘在内的各种功能和流线的叠置。关西机场还实现了与大都市地区的多模式连接，为就业人群提供多种特殊服务和住房，建设临空新城以及防灾基础设施。三种新的室内空间别具一格，既易于辨识，又为乘客提供日常商品，包括："峡谷"——一个集商店、餐厅、休息室以及其他服务休闲设施于一体的开放式多层综合体；"大屋顶"——容纳各种办理客票和登机手续的大厅；以及

discourses that go with them, along with specific territories that became defined, in two senses of the term, and their architectural geographies, again in two senses of that term.

It is probably not possible nor useful to systematically describe all the territories that have emerged during this prolific contemporary period of urban-architectural production. Moreover, it is probably not necessary to do so in order to sufficiently sketch-in and even exemplify the public architecture of East Asian urbanization in this latest, "post-tipping point" stage of modernization. Rather, what I intend to do is to tell six short stories that in effect deal with six kinds of territorial incursion or development. These, in turn, can be subsumed under three commonplace rationales for territorialization in the first place, namely: (1) reaching out to overcome barriers including the isolation of distance; (2) making room much closer to home for new development; and (3) deciding, over time, what to keep from what not to keep, and how to keep it.

I would like to start with the issue of literally overcoming distance – of "flying in and out of town" and the themes of arrival, departure, and linkage, highlighting the territories associated with accomplishing this, ie., airports and all their accoutrements (Fig. 5). Further, the rise of these territories is set against the substantial uptick in East Asian air travel, making it the world's second largest market and with new facilities up at the highest levels of throughput, at around 50 to 70 million passengers per year and higher. Moreover, this is also a short story about the creation and development of a new airport prototype – originally by Aéroports de Paris – not only in scope and scale but also spatial layout, directness, commodiousness of movement, and the organic/naturalistic quality of construction and sub-system integration that went with it.

This new breed of airports begins with Kansai in the early 1990s on its artificially-created off-shore platform, some 40 kilometers southwest of Osaka, Japan. This project was the result of an international competition, won by the Renzo Piano Building Workshop, although it was Aéroports de Paris that essentially programmed and spatially defined the new prototype in response to a request from the transport authority for a critique of the original reference brief. The core idea was deployment of international and domestic flights on different levels of a

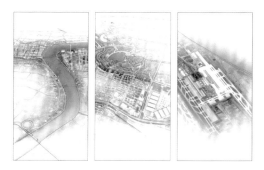

图5 飞入与飞出上海时所见的浦东和浦西
（资料来源：Michael Sypkens）

Fig.5 Flying in and out of town from Pudong to Puxi in Shanghai
(Source: Michael Sypkens)

"翼"——候机厅以及通往不同登机口的通道（图6）。

上海浦东国际机场延续了关西机场的原型。由保罗·安德鲁和巴黎机场公司负责设计的一期工程于1999年启动，2005年开始二期建设。浦东国际机场通过高速公路和磁浮轨道与上海市中心相连。时速450千米的磁浮列车成为中国布局高速铁路网络的早期试验场。除了建筑本身外，浦东国际机场的布局和空间组织也基本沿用了关西机场的模式，并且在多模态地面运输中心及航站楼的分离和集中布局方面进行了优化，规模也更为宏大。论及浦东机场的成功运营，一个关键点在于通过桥梁、隧道以及众多的地铁线路实现了黄浦江两岸便捷的空间联接。这些连接包括1992年建成的南浦大桥、1993年建成的杨浦大桥、2002年建成的卢浦大桥和2005年建成的徐浦大桥，与近20条地铁线路、快速路以及高速路共同构成了大规模的网络布局（图7）。

不过，最为壮观的机场及附属基础设施建设则非香港莫属。一项耗费210亿美元的"机场核心计划"使得香港新机场成为世界最大的单体建筑，或者说至少在施工期间保持了这项纪录。新机场取代了原先更接近市中心的启德机场，并作为1997年香港回归中国的献礼工程（虽然项目到1999年才正式完工）。这项机场核心计划几乎涵盖了从赤鱲角机场到香港中环的所有可能的连接方式，包括10项工程项目，譬如东涌新市镇、北大屿山快速公路和铁路、青马大桥和汲水门大桥，以及采取多模态换乘和香港典型商业模式的九龙站及香港中环站（图8）。赤鱲角机场同关西、浦东机场相比，继续延续了多层交通组织方式，而且严格说来并不存在"国内航班"。此外还有一点略微不同。正如福斯特建筑事务所指出的，设计旨在"回

single multi-floor and multi-use terminal, in lieu of side-by-side arrangements as in the past, thus facilitating advantages of all sorts of overlaps including transfers. At Kansai, multi-modal links to metropolitan areas, the creation of specific services and housing for employment populations were incorporated, including the Rinku New Town and disaster-secured infrastructural supports. Three quite distinctive interior environments also emerged, providing both legibility and commodity to passengers. They were: (1) the "canyon" – an open multi-floored array of shops, restaurants, lounges, and similar services and amenities; (2) the "big roofed" hall for accommodating ticketing and processing; and (3) "the wing" for waiting and movement to specific gates (Fig. 6).

Kansai was then followed by the Pudong International Airport, the first phase of which, under Paul Andreu and Aéroport de Paris, dates from 1999 and the second from 2005. This ensemble is then linked back towards the center of Shanghai by expressways and the experimental Maglev connection to Longyan Road Station near Century Park. At operational speeds of 450 kilometers per hour, this was also an early test bed for high-speed rail links throughout China. The layout at Pudong and other aspects of the spatial organization, if not the architecture per se, follow the model or paradigm established at Kansai, with some refinements such as the separation and centralization of a multi-modal ground transportation center or terminal. The scale was also larger. Also critical to the Pudong operation and, indeed, in realizing the outstanding ambition of physically crossing the Huangpu River from Puxi to Pudong were bridges as well as tunnel crossings, along with extensive subway access. In particular, there were: the Nanpu Bridge, 1992; the Yangpu Bridge, 1993; the Lupu Bridge, 2002; and the Xupu Bridge, 2005. All of this effort was also tied up with the extensive layout and engineering of an eventual 17 to 20-line subway network, and with expressway and highway installations largely above and along existing rights-of-way (Fig. 7).

The grandest airport and infrastructure undertaking, however, occurred in Hong Kong, under the rubric of the Airport Core Programs costing some $21 billion, making it the largest single infrastructure undertaking in the world, at least at the time of construction. Timed to replace the venerable Kai Tak facility

图6 关西机场的"峡谷""大屋顶"及"翼"
（资料来源：Shinkenchiku-Sha）

Fig.6 The "canyon" "big roofed hall" and "wing" at Kansai
(Source: Shinkenchiku-Sha)

图7 连接上海浦西与浦东的桥梁
（资料来源：上海城市建设档案）

Fig.7 Bridging from Puxi to Pudong in Shanghai
(Source: Shanghai Urban Construction Archives)

归简单，放弃传统机场笨重的屋顶形式及其装载的各种设备"，他们利用129个管状模块和网状框架创造出一个轻盈而宁静的空间，很明显是借鉴了之前伦敦斯坦斯特德机场的设计经验。

还有许多机场设施也在这一时期建成，特别是在中国。其中，北京首都国际机场3号航站楼堪称世界上规模最大、跨度最长的建筑，建筑面积为130万平方米，长达3.25千米。此外，还有200年启用的由特里·法雷尔设计的具有"虫状"交通核的仁川国际机场，以及2008年启用的由SOM设计的新加坡3号航站楼，体型更加谦逊却又不失俏皮。其他实例还包括如FOA设计的横滨渡轮码头，中国的各种高铁站，以及结合零售业和文化景观、充斥着艺术装置和表演场所的地铁站综合体。

地域化的另一个问题是如何在物质和文化两方面克服障碍，从而在一个特定场所实现更加全面和完整的城市发展。这个故事讲的是关于上海如何实现从浦西到浦东的跨越。这一雄心壮志至少可以追溯到1919年孙中山在《建国方略》中构想的与外国租界区隔江对峙的"东方大港"的建设愿景。他提出在黄浦江东岸开辟新码头，并开凿运河贯穿浦东。受限于当时资源匮乏和理想过于宏大，这一设想并未付诸实施。但在"二战"后的1946年、1954年和1959年三轮城市总体规划中，浦东的发展趋势已可窥端倪。在正对苏州河河口的黄浦江转弯处尤为明显。但当时局促的财力物力仍无法支撑跨越式发展，俗语道"宁要浦西一张床，不要浦东一间房"，依旧是现实。直到1986年版上海城市总体规划为浦东的发展奠定了基础，特别是1995年的修订稿和1999—2020年上海城市总体规划中提出了卫星城发展计划。更具体而言，1987年确定

closer in to the central city, the project was also wrapped up with the handover of Hong Kong from Britain to China in 1997, even if the program was completed in 1999. With ten specific components, the Airport Core Program touched almost everything, from Chek Lap Kok airport on its island platform at Lantau, all the way back to central Hong Kong. These components included: (1) Tung Chung New Town; (2) North Lantau Expressway and Rail Links; (3) the Tsing Ma and Kap Shui among other extraordinary bridges; and (4) the Kowloon and Hong Kong Central Stations, together with their multi-modal interchanges and extraordinary cross-sections of commercial Hong Kong; etc (Fig. 8). Chek Lap Kok has a slightly different profile than Kansai and Pudong, although the multiple level arrangement remains intact, and domestic flights, strictly speaking, are non-existent. Foster and Partners' aim was, as they put it, "to return to simplicity and to ditch the clunkiness of the conventional airport with its heavy roof laden with equipment". Instead, they created an airy and serene space, based on 129 barrel vault modules with a light lattice frame of diagonal and longitudinal elements. In this, the slightly earlier Stansted airport served as a precedent, evidently.

Many other airport facilities have also been constructed, particularly in China, where there is the Terminal 3 addition to Beijing Capital Airport – the longest and largest building in the world at 3.25 kilometers in length and with a floor area of 1.3 million square meters. In addition, there is also Incheon International with its "insect-like" transportation center by Terry Farrell of 2001, as well as the more modest and yet playful Terminal 3 in Singapore of 2008 by SOM. Also of side-bar interest are other arrival-departure modalities like: (1) Yokohama Ferry Terminal by FOA; (2) China's various High-Speed Train Terminals; and (3) installation of subway stops, doubling as retailing as well as "cultural" landscapes, replete with art installations and performance spaces.

Now, also involved with "reaching out" territorially, as it were, is the overcoming of barriers – both physical and cultural – in order to move more decisively towards fuller and more complete urban development in a particular locale. In this case, the story is about Shanghai and the crossing from Puxi into Pudong, an ambition that dates at least back to Sun Yat Sen's 1919 vision

图8　香港核心机场计划构成要素
（资料来源：Michael Sypkens）

Fig.8　Elements of the Airport Core Program in Hong Kong
(Source: Michael Sypkens)

图9 上海：从陆家嘴到世纪公园
（资料来源：Michael Sypkens）

Fig.9 From Lujiazui to Century Park in Shanghai
(Source: Michael Sypkens)

图10 上海人民广场地区的建筑项目
（资料来源：右上 Jong-Hyn Baek and Pilsoo Maing；
其他 Peter G. Rowe）

Fig.10 The projects of the Renmin precinct in Shanghai
(Source: top right by Jong-Hyn Baek and Pilsoo Maing; others by Peter G. Rowe)

了"浦东新区"的规划原则，1990年政府决策开发开放浦东，1993年浦东新区管理委员会成立。

从浦西到浦东的跨越，塑造了一条贯穿城市的东西向虚轴。对应上述关于地域性的两层含义，这条轴线串联起以下几个各具特色的发展地域：（1）承担文化和行政职能的人民广场地区；（2）承担商业职能的陆家嘴地区；（3）作为轴线的世纪大道；（4）承担浦东新区文化和行政职能的花木地区；（5）作为轴线尽端的世纪公园及其周围地区，承担居住和混合功能，并通过联系东西的交通干线与浦东国际机场相连（图9）。在这些区域内，重要的建筑项目包括位于陆家嘴的三座超高层塔式建筑，即金茂大厦、环球金融中心和尚未竣工的上海中心大厦，位于人民广场的上海大剧院、上海历史博物馆和城市规划展览馆，以及位于花木地区的东方艺术中心、上海浦东新区博物馆、上海市档案馆、上海科技馆和上海国际展览中心等（图10）。

在众多塔式建筑中，1999年竣工的金茂大厦曾一度以88层420.5米的高度成为中国最高的建筑物。它由SOM公司设计，整体形式来自对中国地域文化及建筑传统的诠释，据称是模仿开封塔的形制。101层的上海环球金融中心于2008年竣工，由KPF公司设计，日本东京森大厦株式会社等为主体开发建设。其流线型的外观巧妙地实现了建筑物内部荷载的结构平衡，并据建筑师介绍，其完整的体型旨在凸显"高贵的视觉效果"。尚处于施工进程中的上海中心大厦由Gensler公司设计，具有双层表皮和扭转的外形，预计128层，高达580米。

1998年竣工的上海大剧院由让-马里耶·夏邦杰设计。它采取了与金茂大厦类似的手法，旨在实现现代功

of a "Great Port of the Orient" to rival the Foreign Concessions, with a Bund on the eastern side of the Huangpu and a canal running through Pudong to "cutoff", in effect, the influence of the original Bund. Nothing came of this plan, however, because of a lack of resources and a certain over-ambition. "Post-World War" II aspirations in the direction of Pudong can be seen in the river crossings in the city's 1946, 1954, and 1959 Master Plans. This is especially evident around the bend in the Huangpu opposite the mouth of Suzhou Creek. Yet again, however, little happened due to a lack of resources and the old saying that "it is better to have a hovel in Puxi than a mansion in Pudong" still pertained. Finally, it was the 1986 Shanghai Plan that laid the foundations for Pudong's eventual development, especially as revised in the 1995 Plan and Shanghai Master Plan (1999–2020), with its satellite developments. More specifically, the "Pudong New Area" planning principles were laid down in 1987 followed by an initial scheme – largely under Huang Fuxiang's guidance – in 1990 and an official opening and designation of administrative authority in 1993.

Moving from Puxi to Pudong involved several distinctive territorial developments in both senses of the term, and the creation of a virtual west-east axis across the city. They are: (1) the Renmin Precinct, for cultural and administrative functions; (2) Lujiazui for commercial functions; (3) Century Avenue, which serves as a literal axis; (4) the Huamu District, again for administrative and cultural functions, this time for the Pudong New Area; and (5) Century Park and its environs, terminating the axis, so to speak, hosting a residential and mixed-use environment and picking up the east-west circulatory systems from the Pudong International Airport (Fig. 9). Within these territories, significant architectural projects span from (1) the triplet of three very high-rise towers in Lujiazui, namely the Jin Mao, the World Financial Center, and yet to be realized Shanghai Tower; (2) the Shanghai Grand Theatre, the Heritage Museum, and the Museum of Planning at Renmin; (3) the Oriental Arts Center, the Shanghai-Pudong Museum and Archive, the Science and Technology Museum, and the International Exposition Center at Huamu and beyond (Fig. 10).

Among the towers, the Jin Mao by SOM was completed in 1999 and rises 88 storeys to a height of around 420.5 meters, making it the tallest building in China for a while. In overall form,

能与传统建筑形式之间的平衡。它一方面沿袭了中国传统剧场的上翘式屋顶，以及传统殿堂由"台基""屋身"和"屋顶"形成的三段式构图；另一方面又糅合了经典的德国歌剧院形式，并运用了现代材料和透明幕墙。2003年竣工的东方艺术中心由保罗·安德鲁设计，包含一个约2000座的音乐厅，一个约1000座的歌剧厅和一个330座的演奏厅。所有场馆设施被一个曲面玻璃幕墙所包裹，营造出流动而引人的空间氛围。在花木地区的建筑群设计中，核心要旨是既要符合剧场氛围，又要与周围环境相融合。公园和公共开放空间的建设也响应这一思路，从而弥补了上海中心城区在这方面的严重缺陷，项目包括如延安高速路公园、世纪大道和世纪公园（图11）。

如今，目睹黄浦江东岸这些在短短10~15年内完成的相当杰出和多样化的建筑项目，很明显，从浦西到浦东跨越式发展的世纪梦想已经实现。实际上，这个跨越也显著改变了城市的地域范围和地理格局，新中心之"新"不仅仅在于建成未久，更意味着担负起了新的城市职能。至于其形式和风貌，似乎是可辨识性、宏大感以及场景配置共同作用的结果。在这里，可辨识性的实现来自于清晰的建筑布置、明确的发展计划、基础设施的支持，以及地块的划分与布局。宏大感则体现为大多数建筑物采取一种不受约束的建设方式，拥有属于自己的"喘息空间"。更大的问题在于城市氛围缺乏整体感，如同各个部分的简单拼凑。最后，每个地块都拥有自己独特的场景特征，从陆家嘴高层建筑创造的天际线，到花木地区和人民广场营造的"公园中的建筑"，抑或沿着世纪大道和世纪公园打造的模仿香榭丽舍大街和中央公园的空间环境（图12）。

现在我们开始讨论在离家较近的地区创造发展空间。

it is an inflexion of China's regional culture and building tradition, allegedly modeled after the Kaifeng Pagoda. The Shanghai World Financial Center by KPF of 2008, rises 101 storeys and was developed by the Mori Building Group out of Tokyo Japan. The streamlined appearance, skillfully balancing structural weight within the building section, was also devised, according to the architects, to convey a certain "visual nobility" through its monolithic form. The Shanghai Tower by Gensler is under development and scheduled to rise to some 128 storeys or 580 meters with a double-skinned, torquing form.

Like the Jin Mao Buildling, the Shanghai Grand Theater of 1998 by Jean-Marie Charpentier appears to strike a balance between the traditional architectural expression of the Chinese theaters with their upturned roofs, as well as the traditional Chinese halls with their strong "bases", "middles", and "tops"; while also accommodating the classic German opera house form and contemporary deployment of materials and transparent facades. The Oriental Arts Center by Paul Andreu of 2003 houses a 2,000-seat concert hall, a 1,000-seat drama theater, and a small-scale 330-seat theater dedicated to experimental concerts. All three facilities are then placed within a free-standing, curvilinear glazed outer facade creating a dramatic circulation space. Certainly within the situational logic of its Huamu site, it is the affective aspects of both the theatrical and surrounding spaces that appear uppermost in the complex's design. Parks and public open spaces are also conspicuous in and around this string of developments, making up for severe deficits in these regards in central Shanghai. They include: (1) Yan'an Expressway Park; (2) Century Avenue; and (3) Century Park (Fig. 11).

Now, looking across this rather remarkable and diverse array of projects, all constructed in the brief span of 10–15 years, it is clear that the century-long ambition to cross the Huangpu from Puxi into Pudong has been accomplished. In effect this crossing has also altered appreciably the territorial scope and geography of the city, not the least of which is equipping it with a "new" center in a literal as well as functional sense. As far as form and appearance, the broad features in play seem to involve a mix of legibility, spaciousness, and scenography. Here, legibility courses through in the clear diagrams of building layout and straightforwardness of program, infrastructural support, and the

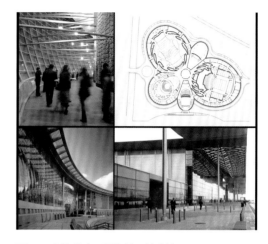

图 11　上海花木及周边地区的建筑项目
（资料来源：左上 Paul Andreu；左下 RTKL；其他 Peter G. Rowe）

Fig.11　Projects in and around Huamu in Shanghai
(Source: top left by Paul Andreu; bottom left by RTKL; others by Peter G. Rowe)

图 12　陆家嘴的高层建筑
（资料来源：Gensler）

Fig.12　The towers and scenography of Lujiazui
(Source: Gensler)

当前的实际工作中涌现出了多种策略。有的是通过土地复垦或填海，以及资产的重新聚集，例如韩国的松岛；有的则是通过对使用条件的重新界定。主要目标大多指向功能的多样化、有计划的住宅供给，以及效率的提升，更具代表性的一点是通过设施建设提升城市竞争力。

在这方面一个突出的实例是新加坡的滨海湾发展计划。该项目在湾口两侧超过360公顷的范围内实施填海工程。它可被视为延续早期殖民时期填海经验的一次最新实践，总计造陆面积达到原有陆地面积的25%。滨海湾开发的重要项目包括：（1）新达城；（2）滨海湾剧场；（3）滨海湾金沙建筑群；（4）最近开放的滨海湾花园。该地区还举办每年一度的一级方程式大奖赛，作为绩效导向下娱乐活动的重要内容（图13）。

建成于1997年的新达城，是滨海湾北区填海开发一期项目中最经久不衰的地标。建筑群包括四座45层的办公大楼、一座小型办公楼、一栋4层的购物中心和新加坡国际会展中心，共同围绕着一个拥有良好景观的广场（图14）。新达城的设计由来自纽约的Tsao and McKown建筑事务所负责，早期的开发目的旨在吸引来自香港的投资，其中的会展中心由香港商家与新加坡政府签约投资建设。滨海艺术中心于2002年落成，由迈克尔·威尔福德事务所设计，是世界上最大的表演艺术综合体之一。它占地6公顷，包含4个剧场，每个剧场专门面向一种特定的表演形式。此外，3层高的购物中心还提供了巨大的商业空间。

这里最为壮观的要数堪称"世外桃源"的滨海湾金沙酒店建筑群，于2010年开业，由摩西·萨夫迪设计。项目最引人注目的是在50层高的酒店塔楼顶部设置了一个非凡悬挑的"天空公园"，提供游泳池和独一无二的俯瞰

lot layout associated with each territory. Spaciousness results from most buildings being sited in an unencumbered manner, with, as it were, their own "breathing room". In more exaggerated circumstances the downside is a somewhat less than fully urban feel to building ensembles and a sense of a whole that does not transcend the sum of its parts. Finally, each territory appears to have its own scenography ranging from the skyline scenography almost certainly derived for Lujiazui on the theme of high-rise building, to the "buildings in a park setting" at Huamu and Renmin, or to the obvious allusions to the Champs Elysée and Central Park along Century Avenue and Century Park(Fig. 12).

Turning now to the idea of making room for new development closer to home, various strategies can be seen at work for making space for contemporary development, either through literal reclamation as shown here at Songdo, South Korea, and re-aggregation of property or through the re-qualification of its use. Further, all in some way were in the service of needed functional diversification, programmatic accommodation, and improved efficiency, typically related to competitive repositioning of cities by way of available facilities.

One very obvious case is Singapore's recent Marina Bay additions, both on the north and to the south, in the form of seaward land reclamations amounting to over 360 hectares in area. In fact, such seaward reclamation is the latest edition of a practice that dates back well into the earlier colonial period. In total, something on the order of 25% of the original land area of the island has been added. Significant within Marina Bay's development are: (1) Suntec City; (2) Esplanade Theaters on the Bay; (3) the Marina Bay Sands complex; and (4) Gardens by the Bay, opened very recently. The area also plays host to the road circuit for the annual Formula One Grand Prix, as part of a performance-oriented entertainment agenda (Fig. 13).

Suntec City, dating from 1997, is the most enduring landmark of Marina Bay North – the first phase of reclamation and development. This complex consists of four 45-storeys office towers, a small office block, a 4-storey podium primarily of retail space, and the "Singapore International Convention and Exhibition Center", all grouped around a well-landscaped plaza (Fig. 14). The architects were Tsao and McKown from New York, and the development was used to attract Hong Kong investment

图13 新加坡的滨海湾发展
（资料来源：Michael Sypkens）

Fig.13 Marina Bay developments in Singapore
(Source: Michael Sypkens)

图14 新加坡滨海湾的新达城
（资料来源：Peter G. Rowe）

Fig.14 Suntec City in Marina Bay, Singapore
(Source: Peter G. Rowe)

整个城市的全景视角（图15）。在新加坡市区重建局的鼓励下，这样的"天空公园"也在新加坡的其他地方出现。该建筑群被标榜为"综合度假胜地"，同时还包含一个赌场，这显然与新加坡早前的严肃作风相违背。但总体而言，如果说能够吸引超过60万游客的到访，这样的设计绝对有其存在的理由。

日本，特别是东京，可以说拥有最丰富的填海造陆类型。自1980年代后期开始，涌现出大批此类规划项目，但多数未能实施，直到新千年房地产泡沫破裂之后，其中的大部分才得以付诸实践。出发点同样是为了提升城市的宜居度、商业吸引力和市场竞争力。

第一种类型是海滨的填海造陆。其理念主要来自丹下健三1960年著名的东京湾计划，尽管从字面意义可以更早地追溯到19世纪后期的明治维新时期。最近的一个实践是1980年代末期到1990年代早期的台场地区的开发，但到目前为止，并未能像预期设想的那样成为商业和娱乐的繁荣之地。

第二种类型是棕地修复。以靠近银座和东京湾的汐留地区铁道用地和车站站场的再开发为例。1987年铁路站场关闭，约22公顷的土地被闲置。之后，东京都政府作为主要业主方经历了很长时间的酝酿，终于在2002—2003年启动更新项目。该项目主要包括数栋办公塔楼，之间以多条不同层面的通道相连接，并且直通东京地铁系统。

第三种类型是产权持有者组成"重建协会"，即所谓的"合作社"，以更加有效地进行物业的再开发，并实现更高的建筑密度和更大的占地面积。以邻近东京中城的六本木新城为例（图16）。作为森大厦株式会社——一个热衷于高层建筑开发的公司——的思想结晶，六本

to the island with the stipulation of a convention-exhibition center being made by the Singaporean government. Esplanade-Theatres on the Bay, completed in 2002 is by Michael Wilford and Partners, and is one of the world's largest performing arts complexes. It sits on a 6-hectare site and houses four theaters, each special purposed for a particular kind of performance. Also included is commercial augmentation via the 3-storey Esplanade Mall.

The spectacular, almost "other worldly" Marina Bay Sands complex opened in 2010 and is by Moshe Safdie. Most conspicuous are the three 50-storey hotel towers crowned by an extraordinary cantilevering "sky park" replete with swimming pool and affording a "one-of-a-kind" panoramic view back over the city (Fig. 15). The "sky park" also crops up elsewhere in Singapore, encouraged by the island state's Urban Redevelopment Authority. Billed as an "integrated resort", the complex contains a gambling casino, clearly a departure programmatically from Singapore's earlier more straight-laced attitude. Overall, the complex if slated to attract 600,000 plus visitors, a significant part of its raison d'etre.

It is in Japan and particularly in Tokyo, however, where perhaps the widest variety of types of territorial reclamation can be seen at work, many planned during the late 1980s although often not implemented until after the bursting of the real estate asset bubble in the present millennium. Again, the rationale was invariably to make the city more amenable, more attractive to commerce, and more competitive.

The first kind of territorial operation is waterfront reclamation that conceptually dates famously back to the 1960 Tokyo Bay Plan of Kenzo Tange, although literally to the Meiji Restoration era in the late 19th Century. More recently are the Daiba or Rainbow Town island developments, from the late 1980s and early 1990s, that, so far, have failed to live up to their earlier billing as productive commercial and recreational venues.

Second, there is the recovery of "brown-field" sites, in this case, rail "rights-of-way" and yards in the area of Shiodome in Minato-ku close to Ginza and to Tokyo Bay. The rail yard was closed in 1987, freeing up about 22 hectares of land. Then, after a long gestation period by the major stakeholder – the Tokyo Metropolitan Government – construction got underway in 2002–2003. The project consists primarily of office towers linked

图15 新加坡滨海湾金沙酒店的快乐时光
（资料来源：Har Ye Kan）

Fig.15 Moments of delight at Marina Bay Sands, Singapore (Source: Har Ye Kan)

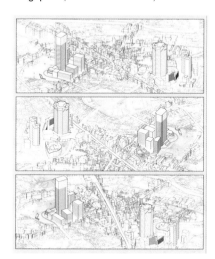

图16 东京六本木新城和中城
（资料来源：Michael Sypkens）

Fig.16 Roppongi Hills and Midtown, Tokyo (Source: Michael Sypkens)

图 17　台北 101 大楼
（资料来源：Michael Sypkens）

Fig.17　Taipei 101 in Xinyi, Taipei
(Source: Michael Sypkens)

图 18　东京表参道时装街
（资料来源：Peter G. Rowe）

Fig.18　Omotesandō Fashion Street in Tokyo
(Source: Peter G. Rowe)

木新城开发的设想始于1980年代，在1990年代完成了主要的规划设计，由KPF公司设计的六本木大厦于2001年开业。产权合并涉及约600名业主，其中的400名仍留在最终的"重建协会"中。整个项目占地11公顷，建筑面积725000平方米，包括商业办公、零售、住宅、酒店、电视演播室，以及大名（领主）庭园的修复工程。该项目获得巨大成功，观光客规模甚至超过了东京的迪士尼公园。

东京中城附近，2008年开始出现第四类的地域再利用实践。当时一处国家政府建筑——日本防卫厅总部——搬离原址，为新的开发空出基地。项目主要的业主是三井公司，由SOM公司进行总体规划。和东亚其它地方的做法类似，项目包含了一系列文化设施，如安藤忠雄与三宅一生联合设计的21-21 Design Sight美术馆，以及隈研吾设计的三得利美术馆等。易道设计公司还在这里设计了一个相对开阔的公园景观。

最后，第五种类型是老旧设施整修与新建设的结合。一个典型案例是东京车站地区的更新。这个项目也是由三井公司负责，自2000年代中期开始。另外值得一提的还有台北信义区的再开发项目。这里坐落着台北101大楼，或称台北金融中心，在当前的纪录保持者迪拜塔未建成时，曾一度为世界最高的建筑（图17）。在日本殖民统治时期，信义区曾经是军队仓库，后来成为一所综合医院的所在地。现在，台北市政厅、孙逸仙博士图书馆以及台北世界贸易中心都坐落于此。虽然遵循了政府的振兴计划，但除去城市荣耀的原因外，如此庞大的建筑物是否有必要实在值得商榷。韩国仁川11千米以外的松岛，也经历了填海造陆的过程，但迄今为止是否成功仍备受质疑，尤其考虑到它是挣扎在对仁川老城的开膛破肚的变革中，制造了

at several levels by pedestrian access, including connection into Tokyo's subway system.

Third, there are the cases of property rights holders forming "redevelopment associations" – so called "Union Pools" – to redevelop property more efficiently meaning at higher density and with larger building footprints. Shown here, is Roppongi Hills, in close proximity to Tokyo Midtown which is described a little later (Fig. 16). Roppongi Hills is the brainchild of Minoru Mori – an ardent believer in high-rise development – and dates back as an idea to the 1980s, although primarily undergoing planning and design in the 1990s, with Roppongi Tower by KPF, opening in 2001. Property consolidation involved some 600 owners, 400 of which remained in the eventual "redevelopment association". All told, the 11 hectare site is occupied by a massive 725,000 square meters of floor space, spread across office commercial, retail commercial, residential, and hotel space, as well as a TV studio by Maki and the restoration of a daimyō (aristocratic) garden. The project has been wildly successful, attracting more visitors than Tokyo Disney World.

Nearby Tokyo Midtown, dating from 2008, involves a fourth kind of territorial reclamation, this time in the decommissioning of a national government site – Japan's Defense Agency Headquarters – to make way for new development. The project was overseen by Mitsui Corporation – the major stakeholder – and masterplanned by SOM. Also included, as seems to be the common recipe these days in East Asia, are cultural facilities like the 21-21 Design Sight exhibition space – a joint venture between Tadao Ando and Issey Miyake – and Kengo Kuma's Suntory Museum of Art. A relatively extensive park landscape was also provided by EDAW.

Finally, a fifth kind of territorial expansion involves relatively conventional refurbishment of older facilities in place along with new construction and the Tokyo Station City area, again by Mitsui, is a conspicuous example, dating again from the mid-2000s. Elsewhere in East Asia, there is the Xinyi District redevelopment in Taipei, which probably deserves mention as the host of Taipei 101 – or the Taipei Financial Center – at one time the tallest building in the world before the Burg Dubai tower, the current record holder (Fig. 17). The Xinyi District occupies what was once an old army depot dating back into the

一个不幸的"双城现象"。

最近东亚城市中还出现了一种关于"街道和时尚展示"的新景象，可以视为"为当代发展创造空间"的特殊形式。以东京的表参道为例，这里没有巴黎圣奥诺雷街或米兰蒙特拿破仑街上优雅庄重的建筑，随处可见的是特别建造和相对小巧的旗舰店、标志性的风貌和精致装扮的时尚路人（图18）。沿着600米的下坡路和400米朝向青山方向的上坡路，排列着各类品牌店铺，包括如：2004年伊东丰雄设计的Tod's店，与街道上榉树的枝杈剪影相映成趣；2003年由赫尔佐格和德梅隆设计的普拉达青山店，拥有珠宝盒状的立面；2003年SANAA设计的迪奥表参道店，双层表皮结构中柔软、半透明的丙烯板仿佛美丽的裙褶一般（图19）。

在服饰和时尚品牌云集的东京银座区，拥有可追溯至17世纪的三越商场，也出现了一些规模较大、设计别致的旗舰店，例如2000年由青木淳设计的路易·威登银座店，2001年由伦佐·皮亚诺建筑工作室设计的爱马仕店，干久美子2004年设计的迪奥银座店，2007年由詹姆斯·卡彭特设计的古奇银座店，等等。最近由坂茂设计的尼古拉斯·哈耶克G中心也落户银座，其中可移动的部分空间供钟表制造商斯沃琪集团使用，同时创造出18米宽的狭长空间，如同垂直布置的传统民居巷道。

虽然没有银座那样悠久的历史，新加坡的乌节路同样值得探讨。在20世纪初乌节路业已成形，自1980年代末期开始进一步成为繁华集市，2005年启动了一项耗资数百万美元的翻修计划（图20）。旨在振兴零售和娱乐业的主要项目包括：（1）乌节中央城；（2）威士马广场；（3）新近开张的爱雍·乌节购物中心。乌节中央城

Japanese colonial period, as well as a hospital complex in more recent times. It is also home to the Taipei Municipal Hall, the Sun Yat-sen Library, and the Taipei World Trade Center. Apart from the "hubris factor", the necessity of such a gargantuan building is less clear, although it certainly complied with Mayors and then Presidents Chen and Ma's city revitalization agendas. Songdo, 11 km outside of Incheon in South Korea, could also be added to the list of seaward reclamations, although now of somewhat dubious success, especially in its evisceration of struggling older Incheon and the creation of an unfortunate "Two-City Phenomenon".

Although differently inflected, a very particular kind of "making room for contemporary development" can be found in the relatively recent emergence of "streets and fashion on display" in East Asia. Here, we can begin with Omotesandō in Tokyo, where instead of the well-mannered buildings of the Rue de Faubourg St Honoré in Paris, or the Milanese palazzi of Via Monte Napoleone, special-built, relatively small, flagship stores have been constructed, replete with iconic exteriors, often alikened to fashion wrapping human bodies (Fig. 18). Along some 600 meters of street on the downward slope plus about 400 meters in the other – Aoyama – direction, one finds, to name but a few outlets: (1) Tod's by Toyo Ito, of 2004, reflecting the branches of the street's majestic zelkova trees; (2) the jewel-box-like facetted facade of Prada Aoyama by Herzog and de Meuron of 2003; and (3) Dior Omotesandō by SANAA of 2003, with its double skin, including the soft, translucent acrylic interior panels like the drape of a dress (Fig. 19).

In the Ginza area of Tokyo, a place of clothing and fashion emporia, like Mitsukoshi dating back into the 17th Century, similar, although slightly larger special-built flagship stores have also emerged. Here, one now finds, for instance, projects like: (1) Louis Vuitton Ginza of 2000 by Jun Aoki; (2) Maison Hermes by the Renzo Piano Building Workshop of 2001; (3) Dior Ginza by Kumiko Inui of 2004; and (4) Gucci Ginza by James Carpenter of 2007. Ginza is also home to Shigeru Ban's more recent Nicolas Hayek G. Center, with its movable parts, somewhat fitting for the Swatch Group of watchmakers, although also creating something of a vertical roji – lane – on the narrow 18 meter-width of the site.

Although not as old as Ginza, Orchard Road in Singapore

图19 东京表参道的普拉达店和迪奥店
（资料来源：Shinkenchiku-Sha）

Fig.19 Prada and Dior at Omotesandō, Tokyo
(Source: Shinkenchiku-Sha)

图20 新加坡乌节路
（资料来源：Michael Sypkens）

Fig.20 Orchard Road, Singapore
(Source: Michael Sypkens)

于2009年开业，购物商场高12层，是城中最高的建筑。"蹦出来"的立面效果反映了城市复兴管理局对于在乌节路的环境中增加"视觉兴奋点"的兴趣。2011年开业的爱雍·乌节购物中心，位于乌节路地铁站上方。多层商场提供了66000平方米的总零售面积，外墙是巨幅的、波浪状的、互动多媒体屏幕，综合体上方矗立着一栋高218米的56层住宅楼（图21）。

类似的还有位于上海的南京路。虽然规模相对小一些，但在中国的商业街发展中却具有举足轻重的地位。在1920—1930年代这段"流金岁月"里，这条通往城外的道路上坐落着先施、新新、大新、永安等各大百货商场。1999年，南京路被改造为步行街，由法国夏邦杰建筑咨询公司和上海著名建筑师、学者郑时龄共同规划。项目建成即大获成功，大批游客蜂拥而来。同时期的还有北京王府井大街项目，尽管相比之下稍显平淡，影响力未及前者。在其它地方也逐渐涌现出一批商业步行街，如武汉的江汉街。

前面提到的第三大类的地域化现象，主要考虑的是随着时间的推移，应对哪些进行保留和再利用，以及如何做到。与其他地方一样，在东亚地区，这个话题涉及历史保护的实践，以及通过环境保护应对现代化带来的负面效应。这两类实践在东亚目前都是相对较新的概念，只有少数尝试，但正受到越来越多的关注。

自1996年开始，由Wood + Zapata建筑事务所设计的上海新天地迅速成为中国针对历史风貌地区进行适应性再利用的明星项目。事实上，项目主要为重建工程，因为原来的里弄建筑实在难以保存，不过早前的"海派"风格在更新后还是得到了合理的再现（图22）。同样在中国，2000年初，《北京旧城25片历史文化保护区保护规

enters the discussion, as a well-shaped street in the early 20th Century and a "trade mart" of significance from the end of the 1980s onwards, becoming the subject of a multi-million dollar facelift in 2005 (Fig. 20). Prominent projects from this revitalization of retail and entertainment venues are: (1) Orchard Central; (2) Wisma Atria; and (3) Ion Orchard, which opened only recently. Orchard Central was opened in 2009, mainly in the form of a 12-storey shopping mall, the tallest in the city. The "pop out" facade also reflects the URA's interest in increasing "visual excitement", as they put it, in Orchard Road's environment. Ion Orchard, opened in 2011, sits above the Orchard Mass Rapid Transit Station. In front of the 66,000 square meters of multi-floor retail, is a street-level curvilinear wave-like multi-media facade. The whole complex is then covered by a 56-storey, 218 meter-high residential tower (Fig. 21).

Similarly, Nanjing Road in Shanghai set the tone – albeit more downscale – in China. This old route out of town was the early home of major department stores in the "Golden Years" of the 1920s and 1930s like the Sincere, Sun Sun, Da Sun, and Wing-On. It was pedestrianized in 1999 with a scheme by Arte Charpentier from France with Zheng Shiling, the prominent Shanghai architect, public servant, and academic. Almost, immediately, it became very successful, especially with visitors to the city. At much the same time, although less iconic or exception in its broad impact, was Wangfujing Dajie in Beijing, to be followed by other pedestrianized streets elsewhere in places like Jianghan Street in Wuhan.

The third broad category of territorialization enumerated earlier, involves deciding, over time, what to keep and re-use and how to do it. In East Asia, as elsewhere, this in turn engages historic preservation and conservation practices, as well as environmental conservation in response to negative externalities of modernization. Both engagements are relatively new to East Asia with some exceptions, although gaining considerable momentum.

Dating from 1996, Xintiandi by Wood and Zapata in Shanghai quickly became one of the "poster projects" for adaptive re-use in China, with historic overtones. In point of fact, it primarily involved re-construction, as most of the former lilong structures were too far gone to preserve, although the earlier ambience

图21　新加坡乌节路的爱雍·乌节购物中心
（资料来源：Har Ye Kan）

Fig.21　Ion Orchard along Orchard Road, Singapore
(Source: Har Ye Kan)

图22　上海新天地的里弄
（资料来源：Peter G. Rowe）

Fig.22　Lanes at Xintiandi, Shanghai
(Source: Peter G. Rowe)

划》的制定显著推进了对旧城的依法保护活动。继而在2007年，上海也颁布实施了《上海市历史文化风貌区和优秀历史建筑保护条例》。除了古迹和宗教建构筑物得到相当的关注外，对于以"胡同"和"里弄"为代表的城市传统街巷的保护也引起了普遍重视。迄今为止，保护方面的努力可以说是成败参半，尽管在经济振兴和再利用方面取得了显著成效。在其他一些地方，比如新加坡的中国广场和远东广场，人们对传统的店屋和"五脚基"骑楼展开抢救，但可能为时已晚。

还有一种更为现代的保护利用方式。在中国的许多地方保存着大量从毛泽东时代起遗留下来的老工业建筑，现在被广泛改造为画廊、工作室、展览厅或表演场所。例如北京的798工厂，前身是1950年代初建造的一家电子厂，在经历了一段几近废弃的时期后，2002—2003年一批艺术家悄悄入驻。现在，它已经被全面合法化了，甚至成为北京双年展的举办地，彰显着国家拥抱当代艺术的态度以及对其市场潜力的热情（图23）。类似的现象在其他地方也屡见不鲜，比如上海，涌现出了一批多产的创意区，包括8号桥、红坊、上海城市雕塑艺术中心、上海1933、M50等。

在日本东京，环境干预存在已久。和西方情况类似，其历史可以追溯到1970年代的环境保护运动。随着现代化进程所导致环境负面影响的日益加剧，更面对环境治理的长期欠账，环境干预也开始在中国出现并加速发展。成都府南河与沙河综合整治项目可以算是相当全面的实施案例。它们分别开始于1990年代中期和2000年左右，在实际工作中综合了防洪、水质提升、污水处理、径流减排、水回收利用甚至交通管理等方面的内容。这两个项目都得

of "haipai" was reasonably well-represented in this updated version (Fig. 22). Staying with China, certainly enactment in Beijing of the 25 Historic Areas and Blocks Plan in early 2000 stimulated a legitimized significant conservation activity in the old city. So did the later Shanghai Historic District Plans and Regulations, implemented in 2007. Apart from monuments and religious structures of considerable concern were the "hutong" and "lilong"lane environments in each city respectively. There, conservation efforts have met with mixed success with the best results to date, as elsewhere, coming in concern with economic revitalization and re-use. Elsewhere, Singapore's China Square and Far East Square, belatedly perhaps, have resulted in some salvaging of some of the city state's shophouse and "five-foot way" tradition.

In a more modern key, widespread use of old industrial buildings, plentiful in many parts of China from the Maoist era, if not before, has taken hold, typically for re-use as galleries, studios, exhibition, and performance spaces of one kind or another. Factory 798 in Beijing on the premises of an electronics factory built in the early 1950s was rather stealthily taken over around 2002–2003 by artists, as it became abandoned. Now, it has become more fully legitimized – even hosting the Beijing Biennale – as State attitudes have warmed to contemporary art and particularly its market potential (Fig. 23). Similar phenomena have occurred in other places, like Shanghai, especially amid the city's prolific array of "creative districts" and places like Bridge 8, Red Town, the Shanghai Sculpture Space, 1933 Shanghai, M50, and so on.

Environmental interventions are of relatively long standing in Japan and Tokyo, dating back to crisis movements in the 1970s, much as in the West. They now are also beginning to happen in China and at an accelerating rate in the face of egregious long-term deferment of the amelioration of negative environmental impacts of modern development. Some are reasonably comprehensive, like the Funanhe and Shahe Revitalization Projects in Chengdu, dating from the mid-1990s and around 2000 respectively. There, clearly at work are combinations of flood control, water quality cleanup, wastewater treatment, runoff mitigation, water recycling, and even traffic management. Both projects have enjoyed significant

图23 北京798艺术空间
（资料来源：Peter G. Rowe）

Fig.23 798 art spaces in Beijing
(Source: Peter G. Rowe)

到了市民和用户的广泛关注和参与，并且包含了活水公园等示范项目，发挥了公众教育的作用。它们如今都成为中国城市最佳实践的案例（图24）。

全面实现历史保护、环境保护和经济振兴相融合的实例，可见于韩国首尔的清溪川复原工程（图25）。此项工程始于2003年，在当时新当选的市长、现在的韩国总统李明博的带领下，旨在使城市重现活力，同时提升市容品质。清溪川的历史可以追溯到朝鲜王朝时期，作为一条自然河川并设有防洪工程，到1960—1970年代朴正熙军政府统治时期被高速公路覆盖而成为现代建设的交通要道。在近年来首尔市中心的一系列复兴项目中，清溪川修复工程发挥了枢纽的作用，串联起其它再开发廊道，包括一条"历史走廊"和光化门广场，以及一条"创意走廊"和东大门设计广场及公园。与清溪川修复工程协同开展的还有一些重要的修复和重建项目，例如在仁寺洞和北村周边地区，这里都有保存完好的地方特色街巷环境。

最后，在更宏观的层面，一些东亚城市体现出历史悠久的、积极的、人造的特质，并持续对当下及未来的发展产生影响。这并非特殊现象，例如当前的东京城市建设正在一定程度上重释江户幕府时代的风格；类似的，中国的一些城市也承袭着帝国时代的建设规制和标准。

事实上，北京就是后者体现在当前城市规划中的一个突出实例，并展现出这种超凡的人为因素对于后续发展的显著影响。北京的城市格局可以追溯到13世纪的元大都城的对称式布局、环形道路和虚拟轴线的组织，之后的历朝历代都留下了发展的印记，但似乎都遵循着某种高度一致的原则。首先，宫城与内城的规整的四边形对称格局由隐含的南北中轴线所控制，依轴线依次布局钟楼、鼓楼、殿

citizen and user input and contained within them demonstration projects, like the Living Water Park, arrived at public education. Both have now also become examples of China's emerging best practices (Fig. 24).

Amalgamation of both historic and environmental conservation, along with economic revitalization, can also be found in the Cheonggyecheon Restoration Project through the center of Seoul, Korea, dating from 2003 (Fig. 25). This was a project spearheaded by Lee Myung-bak, the then newly-elected mayor and now President of the Republic of South Korea, in order to bring life back into the city and to provide additional urban amenity. Historically, Cheonggyecheon was the site of water and flood control improvements dating back into the Joseon Dynasty, as well as used as a right-of-way for the modern improvement, or so it seemed at the time, of a traffic expressway covering over the stream in the 1960s and 1970s under Park Chung-hee's military junta. Stream restoration also served as an armature for other corridors of redevelopment emerging during central Seoul's recent renaissance, including: a "Historic Corridor" and Gwanghwamun Plaza; and a "Creative Corridor" and Dongdaemun Plaza and Park. The project also intersects with significant sites of local restoration and reconstruction around Insadong, for instance, and around Bukchon, with its well-preserved hanok (lane) environments.

Finally, at a more general and macro-scalar level, there is the impact of time-honored, instrumental, and artificial characteristics of some East Asian cities on continuing present and future development. It is not unusual, for example, to speak of the palimpsest quality of Shogunate Edo underlying present-day Tokyo. Nor is it unusual to speak collectively of the persistent regularity and even canonical form of Imperial Chinese cities.

In fact, Beijing is a prominent case of the latter in planning today and a conspicuous example of this almost transcendental artifactual influence over subsequent developments. Dating from the 13th Century and Da Du, the symmetrically-disposed, annular, and virtual axial arrangement of today's Beijing evolved, with each regime leaving their mark, to be sure, but also seeming to march in step, as it were, to a similar drummer, or geomancer. First, the regular quadrangles of the palace city and the inner city became defined with an implied north-south axis with symmetrical

图24　成都沙河地区复兴
（资料来源：Peter G. Rowe）

Fig.24　Shahe revitalization in Chengdu
(Source: Peter G. Rowe)

图25　首尔的清溪川修复
（资料来源：首尔市政府）

Fig.25　Cheonggyecheon restoration in Seoul
(Source: Seoul Metropolitan Government)

堂、门楼等建构筑物。之后，在南部增设了外城。中华人民共和国成立后，又相继形成了东西向的城市轴线，以及不断增扩的环形道路。

到更近一段时期，尤其自1993年国务院批准的修订版北京城市总体规划以来，继续延伸的轴线进一步强化了北京城高度的人工化环境特质。今天我们沿着长安街能够感受到一条强烈的东西向轴线，轴线的一端是CBD，另一端则是金融街和西单。同样也可以看到南北轴线的拓展，从北端的奥林匹克公园到南端北京南站附近的新的开发。根据规划，至少出现了三片新开发地域，分别是：（1）旧城城墙以外东部的CBD；（2）市中心的天安门广场区域；（3）北部的奥林匹克公园。

北京CBD的总体规划方案在2001年由约翰逊·费恩建筑师事务所完成，规划一期建设400~500座新建筑，总计建筑面积1080万平方米，主要用作办公楼和其他商业空间，以及一部分被称作"居家办公"的住宅单元。规划保留了大的街块，建筑形式多样，有独立在基址上的高层建筑，也有填充式的项目，以及沿街拥有一致高度的混合功能体。一个与众不同的例子是CCTV建筑群。它由OMA以及雷姆·库哈斯、奥雷·舍人共同设计，方案于2001年在竞赛中胜出，但由于面临巨大争议，项目直到最近才得以完工。这座巨型设施的总建筑面积超过50万平方米，拥有非同寻常的巨大跨度的悬挑结构，被设计者称为"反摩天楼"。另一个特别的项目是建外SOHO。2002年来自山本理显的设计，这个项目由一组优雅的13~31层高的"居家办公"建筑组成。在这里，基地设计策略体现为一种"场地操作"，包括基地上旋转的组织格网和简洁的景观设计，下沉式的庭院穿插其间，建筑物之间由连廊自由连接（图26）。

alignment along it of the Bell Tower, Drum Tower, Audience Halls, gates, and so on. Then, the outer city was annexed to the south, followed by the east-west axial crossing under the earlier communist regime and the successive annular extensions of ring roads.

In more contemporary times, especially, from the 1993 plan onwards, a further axial influence of Beijing's heightened artifactual quality has emerged. Today, we have the strong East-West axis along Chang'an with the CBD to one side and Financial Street and Xidan on the other. We also have the north-south alignment of the Olympic Green to the north and scheduled new developments in the vicinity of South Station to the south. Within this scheme, at least three territories have also emerged, as both tracts, and as sites of courses of action. They are: (1) the CBD on the East, outside of the old wall of the city; (2) the Tiananmen Precinct in the city center; and (3) the Olympic Green to the north.

The master plan for the CBD was commissioned in 2001 to Johnson Fain Partners, calling, during a first phase, for some 400 to 500 new buildings and 10.8 million square meters of floor space, primarily for office and other commercial space, as well as some residential units in so-called "work-live" arrangements. Block configurations generally remained large with our intended architectural geography varying from tall buildings freestanding on sites, to infill projects as well as mixed-use ensembles of similar rise along street frontages. One exception is the CCTV/TVCC complex by OMA and Rem Koolhaas with Ole Shereen, which was the subject of a competition in 2001, but only completed recently amid considerable controversy. At over 500,000 square meters, the facility is large and unconventionally spans a substantial portion of its site, billed by its authors as an "anti-skyscraper". Another exceptional site strategy was followed at Jian Wai SOHO – a "work-live" ensemble of elegant 13- to 31-storey towers – dating from 2002 by Riken Yamamoto. Here the site strategy is a kind of "field operation" involving an offset organizing grid; as well as a lightly-landscaped plinth, punctuated by sunken courtyards; with loose linkage among structures offered by arcades (Fig. 26).

The Tiananmen precinct is aligned on both the north-south and east-west axes of the city and forms the "axis mundi" of the

图26 北京CCTV和建外SOHO
（资料来源：Shinkenchiku-Sha）

Fig.26 CCTV/TVCC and Jian Wai SOHO in Beijing
(Source: Shinkenchiku-Sha)

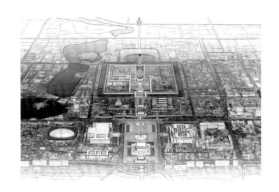

图27 北京天安门广场区域
（资料来源：Michael Sypkens）

Fig.27 The projects and Tiananmen precinct in Beijing
(Source: Michael Sypkens)

图28 北京国家大剧院
（资料来源：Paul Andreu）

Fig.28 Opera House and National Grand Theatre in Beijing
(Source: Paul Andreu)

天安门广场区域座落于城市的南北轴线与东西轴线交汇处，并且形成了共和国时期的"世界核心"（图27）。自1950年代末期以来，这里汇集了一系列重要的公共建筑，北至庄严的天安门城楼，南及古老的箭楼，并与两条地铁线相连接。最近一个值得关注的工程是2007年由保罗·安德鲁设计的国家大剧院（图28）。该项目的三轮竞赛开始于1998年，最终胜出的方案在一个椭圆形的穹顶下容纳了三个表演场地，其中包含一座歌剧院。事实上，随着竞赛的进行，安德鲁的方案同其他竞争者一样，形体看上去更加庞大、完整而抽象。这似乎符合早年赵冬日对于这片地域上建筑体量的最初设想，即"宏伟和壮观"，"整体形态开阔舒展，保持开放性，同时'逐步地'呼应国家未来的事件"（引自1959年）。另一引人瞩目的项目是2000年代中期冯·格康和玛格进行的国家博物馆设计。早期的博物馆由张开济设计，坐落于天安门东侧，作为一座社会主义与现实主义相结合的建筑，被视为1959年"十大建筑"中最杰出的代表。冯·格康和玛格谨慎地尊重了原有建筑的主立面，而对内部空间进行了翻修和更新。

奥林匹克公园，作为值得关注的第三个区域，位于城市的北轴线上，是2008年北京奥运会的主场地。如中国其他大多数公共项目一样，项目设计采取竞赛招标的方式，最终美国Sasaki事务所的方案获得一等奖。方案中顺应城市的主要南北轴线布局了一条次要的轴线，北部尽端是奥林匹克森林公园，作为一个大型的绿色生态保育区（图29）。体育场馆中最引人瞩目的要数奥林匹克体育场，通常被亲切地称为"鸟巢"，由赫尔佐格和德梅隆建筑师事务所设计。体育场的晶格状框架既是结构又是立

Communist regime (Fig. 27). From the late 1950s, it became a site of major public buildings spread across what amounts to a ladder arrangement of blocks flanked by two subway lines, running just to the south of the venerable Tiananmen Gate and to the north of the equally ancient Arrow Tower. One recent project of note is the National Grand Theater of 2007 by Paul Andreu (Fig. 28). Beginning in 1998 with three competition stages, the project accommodates three performance facilities including an opera house under an ellipsoidal dome. Indeed, as the competition wore on, Andreu's scheme, along with those of other competitors, became more seemingly monolithic and expressively abstract. This outcome, in turn, seemed to return to Zhao Dongri's clear original interpretation for the architectural geography of the precinct to be "grand and imposing", as well as "expansive and horizontal in overall configuration, remaining open and, therefore, "progressive" with regard to future events of the nation" (quote from 1959). Another project of note, from the mid-2000s is von Gerkan und Marg's refurbishing of Zhang Kaiji's earlier and imposing National Museum of China – a socialist-realist structure flanking the eastern side of Tiananmen and probably the best, architecturally, of the "Ten Great Buildings of 1959". Appropriately, von Gerkan und Marg respected the main facade of the original building although substantially overhauling and updating its interior.

The Olympic Green – the third territory of note to the north of the city on axis – was, of course, the main site of the 2008 Olympic Games, which Beijing was awarded back in 2001. The subject of a competition like most public projects in China, the venue was planned and designed by Sasaki Associates, including minor axial arrangements in conjunction with the city's main north-south alignment, as well as termination to the north in the Olympic Forest – a substantial green space preserve and conservation area (Fig. 29). Among the sports venues, the Olympic Stadium – affectionately nicknamed the "Bird's Nest" – stands out. Designed by Herzog and de Meuron, the lattice work of the stadium's frame cum facade, introduces an extraordinary sequence of theatrical, almost Piranesian-like spaces around the outside of the arena. Inside, cleverly-tiered seating arrangements allow for accommodation of some 90,000 spectators, but in a manner that feels almost intimate (Fig. 30). The other prominent

面，展示了一种非同寻常的戏剧化效果，外部是皮拉内西式的空间，内部巧妙而紧凑，可以容纳多达9万名观众，却营造出亲切近人的氛围（图30）。其他著名的体育场馆还包括国家游泳中心，俗称"水立方"，由Peddle Thorp & Walker设计。建筑外表皮如同起伏不平的浅蓝色气泡堆积而成，这些加压的ETFE气枕实际上是空间框架结构的组成部分，其空间组织逻辑源自开尔文勋爵在19世纪对三维空间最有效的细分方式的研究。水立方对面，是由朱锫及其都市实践团队设计的北京媒体中心。这是一座外部开窗极少的巨型建筑，整体形式让人联想起"三维条码"。综合体内包含虚拟博物馆、事件档案馆以及数码产品的展览空间。

综上所述，我们探讨的这些项目代表了其所在的东亚城市超越先前经验的建设实践。这些建设主要体现为功能的多样化，并向社会提供了更为广泛的选择空间；同时也折射出民众对于休闲生活的日益广泛而强烈的需求，以及对于环境品质、氛围和舒适度的日趋关注。此外，许多案例也体现为面对之前低效无能的政权所做出的应对策略，尤其聚焦于过度拥挤、环境恶化、服务匮乏和建设不当等问题。整体而言，"竞争力"和"吸引力"都得到了高度关注。

然而，这些故事的背后，至少从西方的角度来看，似乎总有些似曾相识的感觉，或者说是在重复大规模现代化和迈向富裕之前其他地方曾发生过的状况。对很多人而言，全球化带来世界趋同的发展趋势如此之明显，东亚地区是否存在任何例外本身就受到质疑。事实上，这两种看法有可能都具有真实性，至少在表面上如此。不过，它们也可能只是体现在宏观层面关于功能和责任的选择，由此

sports venue is the Aquatic Center – affectionately called the "Water Cube" – with its billowing, blue bubble-like exterior skin. Designed by Peddle Thorp & Walker, the pressurized ETFE pillows were actually integral to a space frame, the spatial organization of which derives from Lord Kelvin's 19th-century speculation of the most efficient way of subdividing a three-dimensional space. Across the road from the Aquatic Center is Digital Beijing by Pei-Zhu and Urbanus. A large and imposing building with little external fenestration and an overall form that conjures up the appearance of a "three-dimensional bar code", the complex houses a virtual museum and archive of the event, as well as exhibition space for digital products.

Now, looking across all this and other material, the projects under discussion invariably represented programmatic expansions beyond earlier experiences in the East Asian host cities. Further, these expansions came mainly by way of functional diversity and presentation of a broader range of choice for those in society. They were also a matter of wider and stronger leisure-time orientations dictated by rising aspirations, as well as by broader concerns with environmental performance, ambience, and amenity. Moreover, in many cases, they were responses to inefficiencies of prior regimes, especially with regard to overcrowding, environmental degradation, poor service and inadequate building. Furthermore, almost across the board, "competitiveness" and "attractiveness" seem to have been strongly equated.

However, lurking behind these narratives for some, at least from a western perspective, there often seems to have been a sense of deja vu, or a repetition of having happened elsewhere before during the broad arc of modernization and rising affluence. For others, the handiwork of globalization's alleged leveling tendencies might also be apparent, raising doubts about any appearance of exceptionalism in the East Asian experience. In fact, there is probably a certain veracity to both these perceptions, at least superficially. Nevertheless, both perceptions might also be simply matters of general subscription to broad categories of functional and amenable choice, although without much by way of concomitant commitment to the same, or even similar urban-architectural outcomes. Many nations, for instance, enjoy football but without the status and styles of play being

图29 北京奥林匹克公园
（资料来源：Michael Sypkens）

Fig.29 The Olympic Green in Beijing
(Source: Michael Sypkens)

图30 北京的"鸟巢"体育场
（资料来源：Ted Lin）

Fig.30 Spaces at the "Bird's Nest" stadium in Beijing
(Source: Ted Lin)

带来的承诺，甚至城市-建筑的产出形式却不尽相同。例如，许多国家热爱足球运动，但状态和打法却相去甚远。

从认可这些是"例外"的视角来看，许多项目的理念、规划和实施都主要依赖于"供应方"，体现为自上而下的指导。即使国家直接资助的程度已不如从前，政府和机构的干预仍然显著存在。从物质空间的层面而言，我们这里探讨的案例通常以大规模的方式并依循其字面含义进行了地域空间的创造或再利用，并且常常采取多项并行的开发形式，比如在滨海湾、浦东、松岛和横滨海湾。

整体上看，与其他地区或其他时期相比，这些实践活动的规模和创新往往令人叹为观止，譬如创造了最高的建筑物（至少曾经一度是）、最大跨度的桥梁、最长的跨海连接、最巨型的建筑物、最快捷的客运、参与人数最多的博览会，以及几乎占全球建设总量一半的项目汇聚于一个地区发生。这些项目中充斥着壮观的景象、精彩的场景、难忘的纪念等等让人拍案叫绝的要素，带来各种欢庆，甚至唤起人们对于过去成功的喜庆。与之形成对比的是，它们的规划方案却往往在风格和定位上相当保守，总体规划平淡无奇，缺乏对于尺度和不同层面细节的精确把握，也几乎看不到公共参与的过程，对于竞争要求以及多元驱动过程几无认知。

如果不考虑那些无法避免的例外和异常，几个共同的特点和趋势已经十分明显，尤其体现在建筑地域风格方面，具体包括如：（1）经常明显地引用范式；（2）常常高度关注工程建造；（3）邀请许多外国建筑师，但会受到来自地方的诸多严格限制；（4）全盘保守，实用主义，善于利用低成本信息。

regarded as anywhere near the same.

Certainly on the "exceptional side" of comparative circumstances, the conceptualization, planning, and implementation of many, if not most, projects have remained on the "supply-side" and has been still primarily top-down in its guidance. Even if not directly state sponsored as in the past, the hand of government authorities and agencies is often clearly present. Physically, the territories under discussion were often created or repurposed in a literal manner and were relatively large. Moreover, they frequently involved extensive "side-by-side" development as at Marina Bay, Pudong, Songdo, and Tokyo or Yokohama Bays.

Seen collectively, the scale and novelty of operations were frequently breathtaking in comparison to other places and at other times: the tallest building, at least for a time; the largest suspension bridge; the longest sea-link; the largest building; the fastest train ride; the most-attended exposition; and almost half of the world's construction going on in one place. Indeed, among projects, spectacle, scenography, monumentality, or some other "wow factor" was rarely ignored, with outcomes that were celebratory, or even triumphant in their evocation as in bygone days elsewhere. By contrast, the plans underlying territorial development tended to be conventional in style and orientation, often involving bland master plans, with a relatively un-nuanced sense of scale or level of detail. Little by way of participatory process, recognition of competing claims or similar pluralistic dynamics seem to have been in play.

Several common traits and tendencies appear to have emerged, particularly in defining the palette of architectural terrain among projects, alongside of inevitable exceptions and even some anomalies. Among these tendencies are: (1) the use often of overt formal references; (2) often a strong convergence with engineering; (3) the presence of many foreign architects but with control very much in the hands of locals; (4) an all-around sense of conservativeness and pragmatically making good use of the low cost of information.

最后，如果说有什么不同的话，那就是在东亚地区所发生的这一切都强化了以下理念，即：从根本上而言，为了我们共同的利益，城市现在和未来都应该是地方性的，采用适合当地的居住方式、目标定位、空间喜好和管理风格。此外，若仔细审视，可以认为"（多样化的）现代主义"似乎是一个更为有效和有吸引力的整体描述。

Finally, if anything, what has happened in East Asia reinforces the idea that, in fundamental ways, cities are and will remain local to all our benefit, by way of inhabitation, orientation, manners of spatial appreciation and styles of management. Moreover, the closeby idea of "modernisms" (plural) seems to be a more useful and attractive overall descriptor the closer one looks.

参考文献 / References

FERGUSON N, 2010. Complexity and Collapse: Empires on the Edge of Chaos[J]. Foreign Affairs. March-April: 18-32.

FOUCAULT M, 1969(2002). The Archaeology of Knowledge[M]. New York: Routledge.

LEVIN M, FORGAN S, HESSLER M, et al., 2010. Urban Modernity: Cultural Innovation in the Second Industrial Revolution[M]. Cambridge, Mass.: MIT Press.

ROWE P, 2005. East Asia Modern: Shaping the Contemporary City[M]. London: Reaktion.

ROWE P, 2011. Emergent Architectural Territories in East Asia[M]. Basel: Birkhaüser.

SENG K, ROWE P, eds. 2004. Shanghai: Architecture and Urbanism of Modern China[M]. Munich: Prestel.

奇迹与地域

MIRACLES AND REGIONS

亚历山大·楚尼斯
代尔夫特技术大学建筑学院荣退教授
清华大学建筑学院客座教授

Alexander Tzonis
Emeritus Professor, Faculty of Architecture, Delft University of Technology
Guest Professor, School of Architecture, Tsinghua University

中文翻译：刘健，朱天禹
Chinese translation by Jian Liu and Tianyu Zhu

人们相信，世界可以被划分为若干"地域"，这些巨大的区域具有独特的环境特点，决定了其居民的身体特性、思维方式和建筑形态；这一观点曾被希腊和罗马的地理学家广为接受，直到最近在西方学者中仍颇为流行。地理学家曾将地域的概念延展到小型地块和中型区域，它们同大型区域一样，具有独特的自然环境、居民群体和人工产物。

在古代的地缘政治学中，大多数中型区域因为清晰可辨的物理屏障而得以被识别；这些物理屏障，诸如山地、峡谷、沙漠、河流和海洋，阻断了地域与外部的联系。地域内居住着不同人群，在空间、社会、文化、经济和技术上受到一种高级权威的管控，通过建立规范日常生活的准则、生产与再生产的关系、食与性的习俗以及管理土地、构筑房屋和房屋综合体的标准，强制或强调居民的凝聚力、排他性和认同感。行使这个权威的管理机构可能是宗教寡头，也可能是民主政权，就像古代雅典或者非洲部落；它通常负责组织一些特殊的仪式和周期性节庆，以加深地域居民对于"内部关联"、共同历史以及共同地域特征的自觉认识。世界各地的人类学和考古学证据已经证实，一个地区聚落的空间设计如何遵循这些理念，又如何通过强化地域的物理屏障——包括空间上的和象征性的，来保持该地域的内部及其组成与外界分开并免受污染，以促进上述理念的建立。这些证据还证实，建筑和城市设计如何在一个地域内，通过规范不同地点之间的位置关系、通达关系和邻近关系——开放的或者是封闭的，对构筑人类交往互动的组织结构及其独有特性发挥重要作用；如何通过建设地标、赋予场地形态、赋予表面颜色、塑造标准的天际线、打磨尺度细腻的装饰细节、操纵光的强度和分布、控制居住空间的声音和嗅觉质量，使得一个地域的内

The belief that the world is divided into "regions", mega-zones characterised by unique environmental conditions that determine the physical constitution of its inhabitants, their way of thinking, and their architecture, was prevalent among Greek and Roman geographers. It continued to be popular among Western authors, till very recently. Geographers applied the concept of region also to smaller land parcels, medium regions, which like the mega regions are also characterized by a unique natural environment, a unique kind of inhabitants and unique artefacts.

Within the geopolitics of the archaic world, most medium regions were recognized by physical barriers that prevented communication with the outside, massive mountain formations, ravines, deserts, rivers, and seas. They were inhabited by human groups under the control - physical, social, cultural, economic, and technological - of a high authority that enforced or stressed the coherence, exclusivity, and identity of the inhabitants through the construction of norms structuring everyday life, relations of production and reproduction, dietary and sexual customs, norms to manage the land, to shape buildings, and building complexes. This governing authority – which could be a religious oligarchy but also democratic as in the case of ancient Athens or African tribes – was usually responsible to develop special rituals and periodic festivals to impress upon the consciousness of the inhabitants of the region a sense of "internal ties", common past, and a common territorial identity. Anthropological and archaeological evidence from all over the world demonstrate how the physical design of the settlements of a region complied with, but also contributed to the construction of these beliefs and the maintenance of their structures through reinforcement of the physical barriers of the region by physical and symbolic barriers keeping the interior of the region and its contents apart from, unpolluted by the outside regions. They also demonstrate how by regulating relations of location, access and contiguity between locations, open and closed, architecture and urban design played a very important role in giving structure and distinctiveness to the organisation of human circulation and interactions within a region. Last of all they show how, by constructing landmarks, giving shape

部环境对其居民而言，变得可辨、熟悉和值得纪念。

人工地区趋于"标准化"的约束条件并非仅仅来自外部强制力。除了人类所具有的群居倾向和建立有利于形成组织化环境的情感联系的倾向之外，自然地理空间向有序的"人工空间"转变也被认为具有同样作用，因为它给人类带来了如下效益：一方面是相对独立的"洁净"疆域的熟悉感和安全感以及由此而来的舒适情感，一种被不了解这个设计的背景和含义的同时代"外来"异地访客理解为美学上的一致性与和谐性的情感；另一方面是基于秩序的人工地区通过相互协调的集体行动，促进了进攻能力、防御能力和生产能力的壮大与提升。

因此，一个地域的形成并不仅仅是其适应当地自然条件的过程，还是将地理上的一片土地转变为一个全新的、某些时候与地理上的自然微气候特性相抵触的人工区域的过程。地域作为人工产物，其演进十分缓慢，造就了多样化的生活环境和生活方式，从文艺复兴起即成为被观察和讨论的对象。然而，尽管历史事实如此，自18世纪以来，一次又一次地，地域的多样性被认为是自然作用的结果，而不是政治力量建设的结果，真正塑造了地域神秘特色的历史起源和演进被逐步弱化直至彻底遗忘。

不可思议的是，这种情况出现在现代科学和理性经验思维不断进步的情况之下，人们对于地域在历史中演进的过去的健忘日甚，而对于"自然"的历史信念却与日俱增，甚至超越了对地域及其居民的历史起源的认识。如果我们将这种新地域主义思想放在现代化进程中来理解，上述悖论即可化解：工业生产带来的资本主义和民族国家的出现需要一种全新的集体认同，以便将城市产业工人以及由来自不同古老地域和不同血统与起源的人所组成的应征

to places, colouring surfaces, configuring standard skylines, forming small scale ornamental details, manipulating the intensity and diffusion of light, but also regulating sound and olfactory quality of inhabited spaces, they made the interior environment of a region for its inhabitants identifiable, familiar, and memorable.

The constraints that "normalised" human-made regions were not imposed by force only. Apart from innate tendencies of humans to form groups and form affective ties that encouraged an organized environment, the transformation of a natural geographical area into an ordered "artefact" was accepted because of the benefits it supplied: on the one hand the comforting feeling of familiarity and security of a set apart, "cleansed" territory – a feeling that a contemporary alien visitor from "outside" who does not know the meaning and context of this design senses as aesthetic congruity and harmony - on the other, the boosting of offensive and defensive power as well as of production that the rule based human-made region facilitated by enabling coordinated collective action.

Thus the overall process of region-making was not only one of adaptation of a region to natural constraints but also of transformation of a piece of geographical land into a new artificial region that several times went against natural, microclimatic idiosyncrasies of geography. Regions as artefacts co-evolved slowly giving birth to the diversity of environments and ways of living observed and discussed since the Renaissance. However, in spite of this historical fact, since the eighteenth century the diversity of regions was considered repeatedly as given by nature and not constructed by political power, the historical origins and evolution that shaped the quasi-mythical identity of these regions being relegated to oblivion.

Strangely this occurred despite the advancement of modern science and rational empirical thinking, the amnesia of the past of the historically evolved regions increasingly promoted together with the historical beliefs in the "natural", beyond history origins of regions and their inhabitants. The paradox is resolved if we see this new regionalist mentality in the context of modernization: the emergence of capitalism driven by industrial production and of the national states both in need of a new

大军聚拢在一起。于是，古老的地域特性被重新利用，成为现代的民族特性，即通过类比古老的地域概念，在更高程度上构建起现代民族国家的概念。

现代世界的发展颠覆了隔绝的地域。它打破了地理政治的障碍，邀请来自不同地域的人们互相交往，把文化、经济、社会和聚居组织的方式从一个地域引入另一个地域，并将它们重新结合和融合，从而使多个不同地域组成的地理空间成为一个全球化的宇宙。

尽管从历史的角度看，全球化推动了资本主义的发展，但它并非资本主义的独有表达，而是与古老的地域共同存在、共同演进。关于共同演进，最有趣的例子之一就是古希腊的地域，它们既保全了唯一性、一致性、"纯粹性"和自治性，同时也并非与世隔绝；它们偶尔成为其他遥远地域网络的组成部分，持续地进行着人员、产品、知识和人工产物的交换互动，结果便是字母、新的法律体系以及我们今天所说的古典建筑的创造性创新发展。

与19世纪的西方不同，国家主义者（经常是种族主义者）认为希腊古典建筑是在一个地域真空中，由纯粹的"阿里乌斯派种族"，即希腊人所独力创造的；然而现代考古学却表明，它其实是初级形态的全球化产物，是许多地中海地域建筑的合成。

另一方面，新生的古典建筑成为一种全球化范式，通过现代全球化的先祖——希腊和罗马帝国，在古代世界传播；之后从18世纪起，又通过殖民消解或吸收既有的地域文化，并通过传播和强加源自帝国都市的标准和规则对这些地域文化进行同化。

然而，如果资本主义没有催生全球化，那么其原有的形态便无法进化；也许没有资本、劳力和知识的动态循环，以

kind of collective identity to bring and keep together industrial urban workers and a conscription army composed of people from different archaic regions, of different stock and roots. Archaic, regional identities were reused turned into modern national identities, the idea of the modern state-nation region constructed to a high degree by analogy to the idea of the archaic region.

The development of the modern world subverted the insular regions. It destroyed geopolitical barriers, inviting people from different regions to interact with each other, importing ways of cultural, economic, social, and settlement organization from one region to another, recombining them, and fusing them turning the geography of multiple diverse regions into one global universe.

Yet, although globalisation assisted the advancement of capitalism historically, it is not an exclusive expression of capitalism. It can be found coexisting and coevolving side by side within the world of archaic regions. One of the most interesting examples of this co-evolution was the case of ancient Greek regions that safeguarded their uniqueness, coherence, "purity", and autonomy but the same time they were not insular. They were part of a network of other, occasionally, distant regions in constant interaction exchanging people, products, knowledge, and artefacts. The result was the development of creative innovations, the alphabet, a new legal system, and what we call today classical architecture.

Thus, contrary to nineteenth century Western, nationalist (very often racist) claims that Greek classical architecture was invented in a regional vacuum single-handedly by one pure "Arian race", the Greeks, modern archaeology however, manifests that it was the product of an early form of globalisation, a synthesis of many regional Mediterranean architectures.

On the other hand, the emergent classical architecture became a global prototype spreading over the ancient world, through the ancestors of modern globalisation, the Hellenistic and Roman empires, and since the 18th century, through colonial ones that dissolved or absorbed existing regions, homogenising them by diffusing and imposing standards and rules originating from the imperial metropolis.

满足繁荣生产和扩大消费的需求，它甚至无法存活。

自1980年代末起，全球化对资本主义扩张的贡献比以往任何时候都更加显著，不断升级的品牌饮料、生活方式和全球明星建筑师的跨境活动，破坏了地域的封闭和孤立，同时也创造了数量空前的财富。

建筑师纷纷涌向建设机遇大量出现，以及近期积累的大量资本准备投资大规模建设的地方，或在那里设立附属机构，例如西部海湾地区、非洲、苏联和中国。有些时候，人们想建房子的地方其实并不需要这些建成空间，之所以建设不过是强烈盼望未来有这样的需求；于是，为了吸引未来的使用者和扩大市场，开发商和房地产经纪人开始面向全球寻找客户。像其他全球产品一样，品牌是最重要的，如同超级制作的电影面对的是全球观众，建筑的形象及其值得记忆的特性以及建筑师的全球声望，都成了最重要的品质，也就是所谓的建筑师"明星"特质，通过文化和技术的完美结合，创造出"名片建筑。

显然，在这样的背景下，地域价值如果不是不受欢迎的，至少也是无关紧要的，因为它们可能抑制项目的全球吸引力。更进一步讲，一个带有自身规则和要求的地域主义建设项目，有可能阻碍全球市场导向的结构所需要的设计自由。

在城市规划中反对地域主义运动的典型是著名建筑师莱姆·库哈斯最近发表的关于"通用城市"的宣言。与地域主义的资源环境保护倾向截然不同，"通用城市"置"保护和发展城市特色"于不顾，致力于建设"没有特性的城市"，"比怀旧版的纽约或巴黎更能准确反映当代都市现实的城市"。

Yet, if capitalism did not invent globalisation, capitalism in its present form could not evolve, perhaps could not even survive, without the dynamic circulation of capital, labour, and knowledge needed on one hand for boosting production on the other for spreading out consumption.

More than ever, globalisation's contribution to the expansion of the capitalist system became evident since the end of the 1980s, through the escalating cross-border move of brand-drinks, life-styles, and global-star architects destroying regional insularity, while creating unprecedented wealth.

Architects moved, or established annexes, where opportunities for building appeared, where heaps of recently accumulated capital searched to invest in massive construction: in the West in the Gulf, Africa, ex-Soviet Union, and China. Several times, there was a call for building in places where built space was not in demand following great expectations that one day it will be needed. To attract such future occupants and enlarge their market, developers and realtors turned to the global clientele. Like any global product, branding was most important. Like a film super-production destined for a global audience, the image and memorability of the scheme became most important, as well as global reputation of the architect, what came to be known "star" quality of the architect delivering "signature architecture" combining cultural as well as technological excellence.

Clearly, under these circumstances, regional values were irrelevant if not undesirable, since they could inhibit the global appeal of the project. In addition, a regionalist building program with its rules and requirements would be obstructing the design freedom that a global market oriented structure needed.

Typical of this anti-regionalist movement in urbanism is a relatively recent manifesto on the "generic city" by a prominent architect, Rem Koolhaas. In contrast to the conservationist tendencies of regionalism, the "generic city" is expected to ignore the values of "preserving and developing a city's identity" aiming at "a city without qualities", "a more accurate reflection of contemporary urban reality than nostalgic visions of New York or Paris".

在中国，此类"全球主义"的建筑比比皆是。在经历了数百年的衰弱、贫穷和发展迟缓后，中国在过去20年里实现了经济腾飞，并参与到全球化进程，投资近万亿进行宏伟的城市建设，"可能是人类历史中最大规模的城市建设运动"。中国急需更多的建成空间，并且根据最新的标准进行建设；它也同样需要开创性项目，以带来新的建造技术，展示中国目前的成就，至少是以全球化方式完成的、面向富裕租户的地产项目（XIAO，2011）。

"面对新的机遇，外国建筑师云集中国"。根据扎哈·哈迪德（Zaha Hadid）的说法，作为回应，（他们中的）许多人将中国视为"一张完美的空白画布"，完全无视地域文脉，在很短的时间内，它们就成为"城市景观中不容置疑的组成部分"（LIU，2011）。

同样的现象也曾出现在19世纪的美国。那时正是美国经济蓬勃发展的时期，建筑师相对匮乏，一些欧洲建筑师来到这里，机械地强制推行"巴黎美术学院派"的设计理念，忽视了地域的传统、限制和可能。

当然，这样的忽视不能全部归罪于建筑师，"城市化进程的'跃进'使得城市盲目发展，忽视了人口控制，只是不切实际地追求经济回报。2010年的一份调查显示，有655个中国城市规划成为'全球城市'……"（GLOBAL TIMES，2011）。

非常普遍的情形是，项目委托方，包括私人企业和公共机构，因为急于展示自己对于现代化和急剧变化的义务和承诺，漠视自身地域的特征，仅仅要求"为了与众不同而建设"。这种情况在2008年危机之前令人眼花缭乱的十年中尤为普遍。

Many of these cases of "globalist" architecture are to be found in China. After hundreds of years of feebleness, poverty, and building inaction, during the last two decades of economic boom China, participating in the globalisation drive, spent about trillion dollars in spectacular construction, "probably the largest urban construction movement in human history." China needed desperately more built space and built according to the latest standards of construction. But it also needed pioneering projects that would bring in new techniques of construction, that would demonstrate China's recent achievements, and last but not least, projects that would compete globally as real estate for wealthy tenants (XIAO, 2011).

"Foreign architects flock to China as new opportunities open up" remarked Xiao Xiangyi in China Daily. Most (of them) respond (ing) to the challenge, in the words of Zaha Hadid, considering China "a perfect blank canvas" —that is ignoring regional context, within a short time they became "an unmistakable part of the urban cityscape" (LIU, 2011).

A similar phenomenon happened in the USA in the 19th century, a period of economic boom for the country, when the country lacked architects. A number of Europeans arrived to impose mechanically "Beaux Arts" schemes disregarding regional traditions, constraints, and possibilities.

One cannot blame the architects for this omission. As professor Xiang Chongling said, "the 'leaping' in urbanization causes cities to develop blindly, ignoring pollution control, and pursuing economic response impractically. A survey result in 2010 showed that 655 Chinese cities are planning on 'going global' ..." (GLOBAL TIMES, 2011)

Very often, especially during the giddy decade before the 2008 crisis, it was those who commissioned the projects, private enterprises or public institutions, who, driven by the urge to manifest their commitment to modernization and radical change, were indifferent to the identity of their region asking for "constructions whose only condition is to stand out from the crowd".

最有说服力的案例之一就是库哈斯设计的新CCTV大楼。他在2002年年末接受委托,计划在2008年奥运会之前完工;他希望这栋大楼不仅要展示全球化和急剧变化已登陆中国,同时也要像为了治疗而插入身体的"针灸"一样,促使长期陷入停顿和衰败的现状传统地域肌理重新焕发活力。

实际上,由单体建筑带动急剧变化的想法并不新鲜,"针灸"的比喻不过是将这种观点神秘化而已。早在1923—1925年,苏联艺术家和建筑师埃尔·利西茨基已经设想到一种新的建筑类型,称其为"云栅",它的形状和CCTV大楼类似,强调水平结构向空中升起,作为功能性和社会性场所来使用。利西茨基设想了八个这样的塔楼,安置在莫斯科环路上的主要路口,作为"先锋艺术"的鼓动器,激发整个城市的革命性变化。同样,CCTV大楼作为一个单体结构,吸引了全球范围的媒体宣传,被《纽约时报》称为"可能是本世纪建造的最伟大的作品";作为解放的代表,它旨在把整个环境从与地域主义密切相关的连续性和保守主义的限制中解放出来(OUROUSSOFF,2011)。

位于西班牙巴斯克地区毕尔巴鄂市的古根海姆博物馆是另一个无视地域的空间特性和文化特性,将全球化建筑嵌入其中,并希望给当地带来新的活力的著名案例。

从19世纪后半叶至1975年,毕尔巴鄂及其所在的巴斯克地区作为工业中心,经历了高水平的发展和繁荣,之后却因传统技术过时、全球化带来生产外迁、1970年代中全球经济衰退以及埃塔(ETA)恐怖组织的影响,很快陷入衰败。在1979年到1985年期间,毕尔巴鄂都市区损失了几乎25%的工业就业岗位(PLAZA(a),2000)。

One of the most expressive examples was the case of the new CCTV building by Rem Koolhaas. He offered the commission end of 2002 and ahead of the 2008 Olympic Games. The building was expected not only to demonstrate that globalisation and radical change arrived in China but also - as "acupuncture needles" function therapeutically inserted in a human body - revive the existing traditional regional tissue that suffered from long periods of stagnation and decline.

Actually, the idea of a single building bringing about radical change is not very original and the metaphor of "acupuncture" only helps mystify the argument. Back in 1923–1925, the Soviet artist and architect El Lissitzki conceived a new building type, he called "cloud-irons" (Wolkenbügel). The shape resembled the CCTV project stressing the potential of raising horizontal structures up in the air to be used as functional and social places. Lissitzki envisages eight such towers, placed on the major intersections of the Boulevard Ring in Moscow, that were going to act as "avant-garde" agitators to incite revolutionary change for the whole city. Similarly, the CCTV single structure, which attracted a unique worldwide publicity, declared by the New York Times as "may be the greatest work of architecture built in this century", an agent of liberation, expected to liberate the whole environment from the constraints of continuity and conservation associated with regionalism (OUROUSSOFF, 2011).

Another renowned case of global architecture inserted in a region ignoring the region's physical and cultural identity but expected to give new life to the region is the Guggenheim museum in Bilbao in the Basque country, Spain.

During the second part of the 19th century until 1975, the city of Bilbao and its Basque Country region enjoyed a high level of development and prosperity as an industrial centre. Since then, due to the obsolescence of the used technology, the migration of the production caused by globalisation, the mid-1970s international recession, and the ETA terrorism, a rapid decline followed. Almost 25% of the industrial jobs in metropolitan Bilbao vanished between 1979 and 1985 (PLAZA (a), 2000).

为了扭转颓势，毕尔巴鄂决定建造一座重要的博物馆，发挥全球文化中心的作用，吸引来自世界各地的游客，就像CCTV大楼一样，利用它的独特性和后现代性，为这座城市及其所在的地区带来复兴的新气象。

恰逢全球化浪潮态势迅猛，毕尔巴鄂决定寻找并引进享有国际声誉的明星机构和明星建筑师。该市的副市长表示，"只有好的建筑是不够的，为了具有吸引力，我们需要响当当的名字"（PONZINI，2010）。

当巴斯克的管理者们在全球范围内寻找可以满足其项目需求的满意人选时，在大西洋的另一面，一座专注于先锋艺术的著名博物馆——在托马斯·克莱恩斯（Tomas Krens）治下的纽约古根海姆博物馆，同样受全球化浪潮的驱使，也在寻找机会改变特性，从一个纽约的地方性机构转化为一个全球化机构。1990年代初，这两种意愿相互碰撞，一拍即合。他们选择的建筑师是弗兰克·盖里，一位享誉全球的明星建筑师，此前与古根海姆有过合作。

不出所料，盖里提出了一个盖里式的"名片建筑"，基于全新的建筑信息模型，确定复杂的几何曲线造型，并开创性地覆以钛合金表皮，整个方案完全不是对地域的资源、潜力和限制条件进行分析的结果。

《纽约时报》对盖里的方案大加赞赏，就像此后赞扬CCTV大楼那样。1997年9月7日的周日刊封面就是这栋建筑的大幅照片；为表达对这一工程神秘力量的敬畏，时任建筑评论员赫尔伯特·穆山穆普在封面照片的文字上称其为"奇迹"，"无法用言语描述的奇迹，其中主要的一个就出现在这里"。这位评论员在文中指出，该项目所在地区"自豪地，如果不是官方地，独立于西班牙其他地区"；在文章结尾，他赞扬道："当我们任何一个人都能

To reverse the decline, Bilbao decided to build a major museum to function as a global cultural centre bringing visitors from all over the world but also, as in the case of the CCTV acupuncture-building, with its individuality and post-modernity, bring about a new spirit of regeneration for the city and the region.

In a fascinating coincidence in the context of roaring globalisation, Bilbao searched to import an internationally famed star-institution and star-architect. The vice-mayor of the city declared, "good architecture is not enough anymore: to seduce we need names" (PONZINI, 2010).

While the Basque administrators searched globally for means to satisfy their project, on the other side of the Atlantic, a famous museum specializing in avant-garde art, the Guggenheim Museum of New York, under the headship of Thomas Krens, driven by the same forces of globalization, searched for an opportunity to alter its identity from a New York institution into a global one. Beginning of the 1990s, the two interests met each other. Found a perfect match. The architect they chose was Frank Gehry, who was already involved with the Guggenheim, a well-known global star-architect.

As expected, Gehry proposed a Gerhy "signature building". Its pioneering Titanium-clad skin was formed according to a sophisticated geometrical curvaceous configuration supported by a new computer program, Building Information Modelling. No surprise, it was not generated following an analysis of the resources, the potentials, and the constraints of the region.

New York Times celebrated Gerhy's scheme as it did with the CCTV. Its architectural critic at that time, Herbert Muschamp, in a text printed on a photo of the building on the cover page of the Sunday Magazine (Sept 7, 1997), in awe of the mystical powers of the project, called it a "miracle": "The word is out that miracles still occur, and that a major one is happening here." In his article, the critic pointed out that the project is located in a region which "proudly, if not officially, independent from the rest of Spain". And the article ended praising the project: "It's a victory for all when any one of us finds a path into freedom, as Frank Gehry has this year in Bilbao." It was enthusiastic but ambiguous conclusion, not to explain which path, whose freedom and from whose tyranny it was referring.

寻找通往自由的道路，就像弗兰克·盖里今年在毕尔巴鄂做到的那样，那就是我们所有人的胜利。"这是一个充满热情却又含糊不清的结论，既没有解释所指的是哪条道路、谁的自由，也没有说明摆脱的又是谁的专制。

考虑到巴斯克地区向马德里争取独立的长期斗争，可以认为上述评论暗示了盖里对巴斯克地区争取自治努力的支持。但实际上，根据1970年年末宪法，该地区要比西班牙其他17个省在文化、财政和政治上享有更多的自治权（PLOGER, 2007）。正如比亚特里兹·普拉扎指出的那样，盖里的工程完全是由"与中央政府资金毫无关联的公共投资"支持的（PLAZA (a), 2000）。

文章或许是在赞美盖里的创新，摆脱了地域文脉、传统和规则的束缚。有趣的是，盖里的客户、古根海姆博物馆项目的委托方——巴斯克政府文化处，一直以来都是由民族党执掌，其职责就是发扬光大巴斯克的地域特色，而从盖里的方案中却很难看出与项目委托方致力于强化地域特色的职责之间的任何关联。实际上，有人曾指出，无视与地域文化的任何联系而将该博物馆置入这一地区的做法只会抑制当地艺术家的创作（BADESCU, 2004; GOMEZ & GONZALEZ, 2001）。

不管怎样，盖里的创作使毕尔巴鄂获益匪浅。1997年10月18日，博物馆由西班牙国王胡安·卡洛斯一世主持开幕，在运营的第一年就创造了接待140万参观者的记录。世界各地的游客蜂拥而至，1998年全年的参观者数量达到1307187人。这个颇为奇特的现象很快就被称为"毕尔巴鄂效应"。

在相当大程度上，导致这一成就的重要因素是古根海姆的品牌效应和宣传机制，以及托马斯·克莱恩斯作为企

Given the long struggle of "País Vasco" for independence from Madrid, one would say the critic suggests Gehry as promoting the autonomist struggles of Basque Country. In fact, the region as a result of the late 1970 Constitution has been enjoying more cultural, fiscal and political autonomy than the 17 provinces of Spain (PLOGER, 2007). Gehry's project, as Beatriz Plaza remarked, was supported by "a public investment made without recourse to central government funds" (PLAZA (a), 2000).

Or was the article praising Gerhy's innovations, freed from regional context, tradition, and rules. Interestingly Gehry's client, responsible with the deal with the Guggenheim Museum, was the Department of Culture of the Basque Government that has always been run by the nationalist party, committed to the promotion of Basque identity. It was difficult to see how much related to the client's commitment to the particularity of the region was Gehry's project. In fact, it has been pointed out that the insertion of the Museum to the region without any connection to the local culture only holds back the activities of the local artists (BADESCU, 2004; GOMEZ & GONZALEZ, 2001).

Nevertheless, Bilbao benefited from Gerhy's work. October 18, 1997, the museum was opened by Juan Carlos I of Spain. The museum attracted a record 1,400,000 visitors during the first year of its operation. Crowds from all over the world rushed to visit the project and a total of 1,307,187 visited the museum in 1998. It was a rather unique phenomenon which soon acquired a name: the "Bilbao effect".

To a great extend, responsible for this success was the Guggenheim as a brand, its publicity machine, and the entrepreneurial genius of Thomas Krens'. However, the real magnet that brought the crowds and the attention of the journalists was the genius of Frank Gehry.

业家的才干。然而，吸引民众以及记者眼球的真正磁石还是弗兰克·盖里的天才创作。

全球媒体对于"毕尔巴鄂效应"的报导红极一时，而几乎所有基于事实、有关该项目对地区影响的评价却姗姗来迟。许多项目聘用外来建筑师的做法引发地域主义者的抗议，例如2000年代末在中国出现的情况。

在中国，当全球主义建设蓬勃发展时，建筑评论却十分少见，这与19世纪末、20世纪初的美国十分相似，直到路易斯·芒福德（Lewis Mumford）出现。在多数人看来，全球化建筑在中国主要城市的出现提供了大量就业岗位，为技术进步和国家富强做出了贡献，而它所产生的地域主义影响却并未引起足够重视。相关的反响出现在新千年的第一个十年结束之际。第一批反对者的批判对象是出现在崇文门附近、具有美国高速公路标牌风格的麦当劳标志（后已拆除）；2007年，针对北京紫禁城内出现的星巴克咖啡店，著名记者芮成钢第一个在网络上发出抗议（WATTS，2007）。此后，约50万人在网络上签署请愿书，几十家报纸报道了这一事件。有趣的是，芮成钢宣称他并不是一个民族主义者，而且他其实很喜欢星巴克咖啡，但他认为把这个全球化的咖啡店植入紫禁城中的做法是对地域–国家象征的"玷污"。

三年以后，这场争论仍在继续。2010年10月15日，时任故宫博物院院长郑欣淼在《人民日报》上表示，"一杯咖啡不会颠覆中华文明"；他把这种反对称为"民族主义"，并且引用卢浮宫作为案例予以反驳，因为卢浮宫里有"各种各样的商店，包括一家中国茶楼"。然而，这个反驳是如此苍白无力，它只能证明全球化的传播和地域文化完整性的消亡不仅仅发生在中国，也发生在法国；尽管

While global journalistic coverage of the "Bilbao effect" was instantaneous, almost all adulatory, evidence based evaluation of the impact of the project on the region was slow to come. However, many projects designed by imported global architects raised regionalist protests, such as those that occurred in China towards the end of the first decade of 2000.

In China, architectural criticism was rare during the globalist building boom, a situation similar to the one in USA at the end of the 19th century and the beginning of the 20th till the arrival of Lewis Mumford. For most people, the appearance of global architecture in the major cities of China was seen as contributing to massive employment, technological advancement, and to the national enrichment. Its regionalist impact attracted less attention. Reactions emerged towards the end of the first decade of the millennium. One of the first retorts was aimed against the appearance of an American highway-style McDonald sign, now gone, in front of the Chongwenmen Gate . Later, in 2007, regarding the arrival of a Starbucks coffee shop inside Beijing's Forbidden City, the first to protest online was the prominent journalist Rui Chenggang (WATTS, 2007). Following that, half a million people signed an online petition and dozens of newspapers carried prominent stories. Interestingly, Rui declared that he is not a nationalist and in fact that he likes the Starbucks coffee. But he called the insertion of the global coffee shop within the regional-national symbolic structure "contamination".

Three years later the debate continued and October 15, 2010 in People's Daily, Zheng Xinmiao, president of China's Palace Museum, declared that "a cup of coffee cannot overthrow Chinese civilization". He called the reaction "nationalist" and in his defense he cited the case of the Louvre which contained "various forms of shops including a Chinese teahouse". It was a week defense that only confirmed the spread of globalization and the demise of regional cultural integrity not only in China but also in France. After all even the Louvre from a historical point of view was the product of a unique regional culture, on its way to become an instrument of a raising global empire.

从历史的角度看，卢浮宫本身是一个独一无二的地域文化产物，但它最终却沦为正在崛起的全球化帝国的工具。

库哈斯的CCTV大楼的影响则更多地体现在视觉干扰上——因为它的尺度和形态，其次才是文化意义上。这个设计不仅怠慢了中国城市的地域传统，而且"通过失礼地模仿拳击短裤侮辱了中国人民"；它就像降落在地域肌理中的外来物，并且这种感受没有随时光流逝而淡去。

全球化建筑有违地域特性的连贯，并会带来严重经济恶果，安亭新镇的失败便是其中的案例之一；它没能像计划中那样吸引居民，反而成了一座鬼城。按照规划，安亭是位于上海郊区的一个大型居住区，可容纳居民50000人；它同时也是一个充满活力的工业区，主要生产德国汽车和汽车配件。项目指定的设计师是著名的德国建筑师阿尔伯特·施皮尔。在当时的中国，复制欧洲城市是一个非常成功的建筑时尚，有些此类项目就在上海近郊，而最有名的则是建在中国南部的哈施塔特小镇，基本完整复制了奥地利的豪斯塔特村；因此，安亭新镇的开发商要求施皮尔模仿一座德国的传统城镇，把安亭设计成为一个德式风格的迷你小镇。施皮尔本人并不想为某种意象和宣传而设计，他有另外的想法，即建造一个能"代表现代的、环境友好的德国，拥有双层玻璃和中央供暖，而不是黑森林式浪漫"的居住之地。最终结果就像杨希凡（音译）在《明镜》周刊网络版上报道的那样，安亭成为一座"德国城镇……像是斯图加特或者卡塞尔的新住区，到处都是包豪斯功能主义风格的3~5层楼房"（YANG, 2011）。

很难解释安亭新镇失败的原因。事实上，开发商和地区政府并未如约为新镇提供必要的服务设施；但是真正击退客户的还是建筑和城市设计，它既没有像哈施塔特那样

Culturally less significant but visually more disturbing, because of its scale and configuration, was the new CCTV building by Rem Koolhaas. Liu Yujie, in the above article argued that the project not only snubbed regional traditions of Chinese urbanism but also "humiliated the Chinese by its irreverent resemblance to a pair of boxer shorts". As years went by the feelings that the project was an alien intervention into the regional tissue did not diminish.

Among the cases of global architecture violating regional coherence had serious economic repercussions was the failure of Anting to attract occupants turning the project into a ghost town. Anting is a large residential quarter planned for 50,000 people built outside Shanghai, a vibrant industrial area identified to a high degree with the production of German cars or car components. The chosen designer was the prominent German architect, Albert Speer. Following a highly successful contemporary vogue in China copying European cities, some in the vicinity of Shanghai – the most well known the Hashitate in southern China, an exact copy of the Austrian village of Hallstatt - the Chinese developers asked Speer to design a German mini town resembling a German traditional town. Speer, who was not inclined to design projects for the sake of imageability and publicity, had another idea: to produce a settlement that "represented modern, environmentally-friendly Germany, with double-glazing and central heating instead of Black Forest romance". The result was, in the words of Xifan Yang on Spiegel on Line a "German Town … like a new residential district in Stuttgart or Kassel, with buildings three to five storeys high in the functional Bauhaus style" (YANG, 2011).

The failure is difficult to explain. The fact is that the developers and the regional administration did not supply the promised necessary services for the project. But what repulses clients is the very architectural and urban design which neither succeeded to offer a touristic illusion as Hashitate did, nor an authentic response to the regional context which was not only an architectural "style" but an urban design and landscape tissue and a way of life.

成功塑造旅游景象，也没能真正回应地域文脉。地域文化不仅仅是一种建筑"风格"，更是城市设计、景观肌理和生活方式。

随着地域主义对全球化建筑入侵中国城市的抵制日渐增强，一些建筑师决定努力展示他们的"中国性"和地域特色；他们在新的结构中引入一些形式化的空间元素，以使建筑轮廓能够模仿一些重要的但却毫不相关的中国著名历史建筑。

《中国日报》曾报道，一栋新的高层建筑将包含"许多传统文化元素"，它将拥有孔明灯般的形态，"唤起人们对古代中国城市的记忆"（ZHENG, 2011）！开发商表示，因为这座位于北京的新摩天大楼的形态来自一件古代的礼器，因此它将被命名为中国樽（CHINA DAILY, 2011）。

由于房屋和城市对日常生活的影响如此广泛，因此建筑与都市主义成为抨击失败的全球化的主要焦点之一。然而，在当今建筑和城市的全球化实践令人失望的背后，则有更加深层次的思想认识原因。

2012年1月在瑞士小镇达沃斯举办的世界经济论坛给出了问题的答案。达沃斯论坛汇集了来自政府、经济、创意企业、学术领域的主要世界领袖，当然也包括了明星建筑师；作为最受欢迎的国际论坛，它的创立就是为了在世界范围内打破边界束缚、推动"自由"流动。它也是被萨缪尔·亨廷顿统称为"达沃斯人"的活动平台，他们是跨国公司的拥趸，每年都在达沃斯聚会，庆祝全球化所取得的空前财富。论坛组织者出于对地缘政治趋势的敏感，体会到全球化所面临的最新张力和抵制，希望能从全球化角度理解地域选择，并且意识到重新探索某种形式的地域主义的必要性，因此将2012年的论坛主题确定

As the regionalist reactions against the intrusion of global architecture in Chinese cities became more numerous, several architects decided to try to demonstrate their "Chineseness" and regional identity introducing in their new structures formal spatial elements from making the silhouette of their buildings resemble important but otherwise irrelevant, famous historical Chinese artefacts.

The China Daily reported that the new tall building will include "many elements from traditional culture' which will 'remind people of the ancient Chinese city". It will have the shape of Kongming Lamp (ZHENG, 2011)! The developer stated that design of the new skyscraper in Beijing is drawing from an ancient ritual vessel and for this reason the building is named Zhongguo Zun, meaning China Goblet (CHINA DAILY, 2011).

Because buildings and cities affect so much everyday life, architecture and urbanism became one of the main focuses of attacks against failed globalisation. However, behind the architectural and urban let-downs of globalisation as practiced today there were deeper problems of mentality.

Revealing is the case of the latest meeting of the World Economic Forum in the small Swiss town Davos in January 2012. The Davos Forum, the most popular gathering of major world-leaders from governments, corporations, creative enterprises, academia, as well as star architects, was founded to promote "free" movement around the world and the elimination of barriers. It has been also a platform of the "Davos Man", a generic name given by Samuel Huntington for the multinational faithful who congregated at Davos annually, to celebrate the unprecedented wealth globalisation achieved. The same time, the organizers, sensitive to shifting geopolitical trends, in recognition of the latest strains and opposition faced by the dominance of globalisation, in the hope to develop a global understanding of the regional alternative and in recognition of the need to re-explore some form of regionalism dedicated 2012 meeting to the theme of "The Rise of Regions in a G-Zero World" (BREMMER (a), 2012; BREMMER (b), 2012).

为"G-0世界中的地域崛起"（BREMMER (a), 2012; BREMMER (b), 2012）。

然而出乎组织者的意料，2012年的论坛显然缺少了某些内容。经过仔细查看发现，除了少量特殊情况之外，与往年不同的是，中国高级官员并未出现在1月24日的大会上！又经过一段时间的思考才发现，2012年达沃斯论坛与"地域性的"中国新年节日在时间上冲突了（ZHANG, 2012）。

尽管这不过是组织者的官僚疏忽，但它却传递出一条令人尴尬的信息：世界经济论坛作为当代致力于全球化的重要机构，即使年会议程讨论的主题就是"地域"问题，仍然没有做好准备，接受继而尊重不同地域之间的深层文化差异。当然，或许还存在一个更加令人尴尬的言外之意，那就是在许多全球化运动背后，经常存在一个潜在的规则，即以牺牲一个地域的价值为代价来强化另一个地域的价值。

与这些从地域主义视角出发对全球化进程进行批判的做法截然不同，全世界的记者们继续盛赞毕尔巴鄂博物馆传奇般的胜利，是带动低迷地区重生和打造新型住区的"范例"。甚至在工程完工之前，世界各地就有许多工程计划模仿毕尔巴鄂的做法，确信"毕尔巴鄂效应"会同样奏效。

古根海姆基金会主席宣称，在毕尔巴鄂之后，他收到了大约60份提议，要求参与世界各地的城市发展项目（PONZINI, 2010）。阿布扎比、瓜达拉哈拉、维尔纽斯和达拉斯艺术区也宣布了类似的项目计划。

显然，针对盖里的试验缺乏基于事实的合理评价，特别是对其地域影响缺乏基于事实的合理评价，但这似乎并未影响这类工程的涌现，只是因为2008年经济危机有所缩减。

另一方面，这种大规模的积极宣传并未阻止研究者收

But much to the surprise of the organizers, the 2012 meeting appeared to be missing something. Under closer scrutiny it became clear that with few exceptions, in contrast to previous years, on January 24, the senior Chinese officials had not come! It took some reflection to discover that the 2012 Davos event collided with … the festivities of the "regional" Chinese New Year (ZHANG, 2012).

Even if that was a simple bureaucratic neglect on the side of the organizers, it gave away a clear embarrassing message that even if the question of "regions" was the main theme of the program of its annual meeting, the World Economic Forum, a leading institution of our time, dedicated to the case of globalisation, was not yet prepared to accept, and consequently to respect, the existence of deep cultural differences between regions. Perhaps there was also one more disconcerting implication, that very often, behind many campaigns for globalisation is an underlying agenda to promote the values of one region at the expense of the values of another.

In contrast to these criticisms of globalisation carried out from a regionalist point of view, journalists around the world continued to acknowledge the Bilbao Museum as a legendary triumph leading to the canonization of the project as a "paradigm" for regeneration of stagnant regions as well as for producing new settlements. Accordingly, even before the completion of the building, many projects around the world were planned imitating the Bilbao process with the conviction that the "Bilbao effect" will follow.

The director of the Guggenheim Foundation claimed that, after the Bilbao experience, he received about sixty proposals for participation in urban development projects in the world (PONZINI, 2010). Similar undertakings were announced for Abu Dhabi, Guadalajara, Vilnius, and Dallas Arts District.

It seems that the absence of proper evidence-based assessment of the Gerhy experiment, especially of its regional impact, did not influence the proliferation of such projects, to be curtailed only with the economic crisis of 2008.

集和分析相关数据，揭示这项工程的真正影响，包括它为地区带来的广泛的社会和经济影响，而不仅仅是博物馆参观者的数量。正如前文断言，毕尔巴鄂已经成为一个杰出的实证案例（PLAZA (a), 2000）。

关于"毕尔巴鄂效应"，研究者努力回答的主要问题之一就是，毕尔巴鄂城市复兴的成功是这座博物馆带来的结果，还是有其他因素发挥了同样或者更大的作用。分析结果表明，盖里的博物馆建设并非地区政府和毕尔巴鄂市为扭转地区衰退而采取的孤立行动；地区复兴计划是一个综合计划，包括了新的《都市规划1989》《都市区空间规划1994》《都市区复兴战略规划1992》，以及一系列交通和基础设施工程，例如由圣地亚哥·卡拉特拉瓦设计的新机场、由米歇尔·维尔福德和詹姆斯·施特林设计的交通中心，以及由西萨·佩里设计的大片滨水地区开发。巴斯克地区大学经济学院经济政策教授比亚特里兹·普拉扎如此总结道："毕尔巴鄂是一个庞大的、连贯的公共政策整体的组成部分。"（PLAZA (b), 2008）如果没有这些总体规划和城市设计项目，"毕尔巴鄂效应"能否发生值得怀疑（PLAZA (a), 2000）。

通过分析有关毕尔巴鄂博物馆案例的相关证据，巴斯克地区大学应用经济学副教授阿朗特克萨·罗德里格兹（Arantxa Rodriguez）也得出类似结论："将毕尔巴鄂的复兴全部归功于盖里的建筑，就像媒体在它开业第一年所做的那样，是错误的。毕尔巴鄂的城市复兴是一个在不同尺度上一系列动态变化的多元进程……显然超越了将其简单归结为'古根海姆效应'的范畴。"（RODRIGUEZ, 2011）。

除了在旅游商业领域的成就外，这项工程对地区经济

On the other hand, this massive positive publicity did not deter researchers from accumulating and analysing data of the real impact of the project, taking into account its social and broader economic influence on the region, rather than reporting the figures of visitors to the museum only. As stated, Bilbao became an outstanding test case (PLAZA (a), 2000).

One of the key questions researchers tried to answer concerning the "Bilbao effect" was if the success of Bilbao's urban regeneration was the result of the museum as such or if other factors played an equal if not greater role. The result of the inquiry showed that the construction of the Gehry Museum was not an isolated action undertaken by the regional administration and the city of Bilbao to reverse the decline of the area. The revitalization program encompassed of a comprehensive plan that included the new *Municipal Urban Plan* (1989), the *Metropolitan Spatial Plan* (1994) and the *Strategic Plan for Metropolitan Revitalization* (1992), as well as a system of transport and infrastructure projects, such as the new airport by Santiago Calatrava, a transportation centre by Michael Wilford and James Stirling, and a vast waterfront development designed by Cesar Pelli. As Beatriz Plaza, professor of economic policy in the Faculty of Economics at the University of the Basque Country concluded, "Bilbao is an integral part of a larger coherent public policy." (PLAZA (b), 2008) Without these major planning and urban design projects it is doubtful if the "Bilbao effect" would have happened (PLAZA (a), 2000).

Looking into the evidence accumulated on the Bilbao Museum case, Arantxa Rodriguez, Associate Professor of Applied Economics at the University of the Basque Country in Bilbao, arrived to similar conclusions: "Crediting the regeneration of Bilbao, as the press did during the first years of its operation, to the Gehry building is erroneous. The urban regeneration in Bilbao is a multidimensional process involving a series of dynamics at various scales ... that extend well beyond the simplistic reduction of 'the Guggenheim effect'." (RODRIGUEZ, 2011)

Another issue examined apart from the touristic-commercial achievement of the insertion of the project was its economic impact on the region. As Arantxa Rodriguez argued, contrary

的影响是被考察的另一个方面。阿朗特克萨·罗德里格兹认为，与预期相反，古根海姆博物馆的"国际资本投资和对城市的主导作用十分有限"；尽管曾在全球范围内进行了大力宣传，"吸引直接外国（国际）投资以及对城市发挥主导作用的预期并未实现"。

该项工程对地区文化和社会的影响也同样引起研究者的兴趣。古根海姆未像预期那样，对地区文化产生"推进"作用，反而起到了抑制作用；它对地区的社会影响也是消极的，博物馆建设促成了"社会和空间的排斥，而这种不均衡再开发可能加剧都市区内已有的社会空间和功能空间的分异"。这还没有考虑不均衡再开发可能带来的其他潜在影响，例如全球化进一步恶化而不是改善地区的社会质量（RODRIGUEZ，2011）。

最有趣的观察之一是年轻学者格鲁依阿·巴迪斯库利用多种标准，对毕尔巴鄂的地区效应进行的分析（BADESCU，2009）。他选择了比较案例研究作为方法论工具，将毕尔巴鄂和英国北部城市谢菲尔德进行对比考察；结果显示，比较研究的结论比单一对象研究更具普遍性，适用范围也更加广泛。毕尔巴鄂和谢菲尔德大约在同一时期开始钢铁生产，并且因为煤矿产业而成为地区中心；同样大约是在同一时期，即1970到1980年代，它们的生产开始变得落后，快速发展也逐渐被严重衰退所取代；为了走出困境，两个城市在1990年代开始重视文化政策和旅游产业，努力促进复兴发展。

在社会层面上，巴迪斯库观察到两个城市的不同之处。如前文所述，毕尔巴鄂狭隘地将重点放在单一的明星建筑上，造成了社会的排斥和极化；与此相反，谢菲尔德采取了社会驱动政策和基于社区的建设项目，"为边缘群

to expectations, the Guggenheim Museum, the "international capital investments and key command functions, are very limited". Despite the strong global advertising campaign, "the original expectations regarding the attraction of direct foreign (international) investment and command functions to the city have not been met".

Researchers were also interested in the cultural and social impact of the project on the region. The Guggenheim effect on regional culture was not "propulsive" as expected but rather repressive. Equally negative was its social influence on the region. The Museum encouraged "social and spatial exclusion as uneven redevelopment may intensify existing social and functional divisions of space within the metropolitan area". The implications of uneven redevelopment were not taken into consideration, globalization worsening rather than improving the social quality of the region (RODRIGUEZ, 2011).

One of the most interesting investigations about the Bilbao effect on the region considering multiple criteria was carried out by a young researcher, Gruia Bădescu (BADESCU, 2009). Bădescu decided to choose an interesting methodological tool, the comparative case study. As a result, the conclusions of the comparative inquiry are more generalizable than the study of a single object and they have a broader applicability. He examined side by side Bilbao and Sheffield a northern English city. Both Bilbao and Sheffield began producing steel and became thriving regional centres for coal mining industry at about the same period. And at about the same time, between 1970s and 1980s, their production became obsolete and their rapid growth was succeeded by severe decline. To overcome the crisis, in the 1990s, both cities undertook campaigns focusing on cultural policies and tourism as means for regeneration.

Considering the results according to a social point of view, Bădescu observed that the two cities diverged. As opposed to Bilbao's narrow focus on a single star-building that created, as mentioned above, social exclusion and polarization, Sheffield followed socially- driven policies and community based projects "providing members of marginal groups employment, job training, and inclusion" that resulted in increased social cohesion.

体成员提供就业岗位、职业培训和包容",最终提高了社会凝聚力。

此外,巴迪斯库还引用经验研究的结论,指出相对毕尔巴鄂,谢菲尔德对城市危机的社会因素关注较少(HOLMES & BEEBANJAN, 2007;PLOGER, 2007),而是注重文化项目和文化建筑,将其视为媒介,在更大尺度上形成包容的环境,而且也没有利用任何明星建筑。尽管毕尔巴鄂将大量资金投入基础设施和诸多房屋建设,它的城市复兴却主要是由单一明星建筑产生的神秘隐喻力量所驱动的。因此,谢菲尔德更接近以地域的观点思考聚落肌理的生态、经济、社会和文化特性,近期的报告也表明至少到目前,这种方法是有效的。

从毕尔巴鄂以及上文讨论的其他案例中,我们可以看到,开发商、都市主义者和政府利用具有诱惑性但也有误导性的比喻,诸如"针灸"或者"毕尔巴鄂效应",将全球化明星建筑引进并且植入城市和区域之中,抑制了地域在社会、经济和文化方面的条件和潜力。其结果是,引进的明星解决方案,无论其作为个体多么出色,像报道的那样在某一方面立刻成为全球奇迹,但从长远来看,考虑到环境质量的多元因素,它们可能无法带来真正的地区复兴。

我们在文章开篇曾提到封闭、纯粹、静止的地域,这种陈旧的观念已经终结。表面化的、怀旧风的和文体上的碎片不能使它复生,新的边界也无法使它继续沟通和交流。然而,在富有"创造力"但同时也富有"破坏力"的全球化动态发展背景下,地域概念作为生态、经济、社会和文化方面的组织工具,用以创造或者重构和谐的、可持续的环境肌理,今时更胜以往。

In addition, in contrast to Bilbao, less attention to the social aspect of the urban crisis, Bădescu remarked, citing results of empirical studies (HOLMES & BEEBANJAN, 2007; PLOGER, 2007). Sheffield focused on cultural projects and buildings as means to generate inclusive spaces on a broader scale and employed no star-architecture. Despite the significant investment of Bilbao in infrastructure and a cluster of buildings its regeneration was driven primarily by the mystical metaphor of the generative power of a single star-building. Sheffield came closer to thinking regionally about the future ecological, economic, social, and cultural of the settlement as a tissue. The recent reports announce that or the moment it works.

We can conclude from the Bilbao case and from the cases we discussed above that developers, urbanists, and administrators, using seductive but misleading metaphors such as "acupuncture" or "Bilbao effect", imported and inserted global star-architecture in cities and regions suppressing social, economic, and cultural regional constraints and potentials. As a result, imported star-solutions whatever their excellence as individual objects maybe reported as instant one-dimension global miracles, but in the long run taking into account multiple dimensions of environmental quality they may fail to bring about realistic regional regeneration.

The archaic idea of a closed purified static region, we mentioned in the beginning of this text, is over and done. Neither cosmetic nostalgic stylistic fragments can bring it back nor new barriers to communication and exchange. However, given the "creative" but still "destructive" development of dynamic globalization, the concept of region as an organizing ecological, economic, social, and cultural tool to recreate or create harmonious sustainable environmental tissue is more contemporary than any time before.

参考文献 / References

BĂDESCU G, 2009. Steel-Town Makeover: Evaluating Urban Regeneration Policy in Sheffield and Bilbao[M]. European city seminars.

BREMMER I (a), 2012. Davos 2012: The Rise of Regions in a G-Zero World[N]. The Atlantic. 2012-1-24.

BREMMER I (b), 2012. Decline of global institutions means we best embrace regionalism. Financial Times. 2012-1-27.

Davide Ponzini. Bilbao effects and narrative defects. Cahiers de recherche du Programme Villes & territoires. n° 2010.

Global Times, 2011. China's rapid urbanization should follow objective law of cities' development[N]. 2011-2-14.

GOMEZ M, GONZALEZ S, 2001. A reply to Beatriz Plaza's "The Guggenheim-Bilbao Museum Effect" [J]. International Journal of Urban and Regional Research, 25 (4): 898-900.

HOLMES K, BEEBANJAN Y, 2007. City centre master planning and cultural spaces: A case study of Sheffield[J]. Journal of Retail and Leisure Property, 6: 29-46.

LIU Y J, 2011. Bad boy architects & China's new face[N]. China Daily. 2011-10-16.

OUROUSSOFF N, 2011. Koolhaas, Delirious in Beijing[N]. New York Times. 2011-7-11.

PLAZA B (a), 2000. Evaluating the Influence of a Large Cultural Artifact in the Attraction of Tourism, The Guggenheim Museum Bilbao Case[J]. Urban Affairs Review, 36: 264-274.

PLAZA B (b), 2008. On some Challenges and Conditions for the Guggenheim Museum Bilbao to be an Effective Economic Re-activator. International Journal of Urban and Regional Research, 32(2): 506-551.

PLÖGER J, 2007. Bilbao city report[R]. Case report 43. LSE, CASE.

RODRIGUEZ A, 2011. Accrediting the success of Bilbao's urban regeneration to the Guggenheim Museum is misguided[M]. European Urban Knowledge Network. 2011-1-28.

WATTS J, 2007. Starbucks faces eviction from the Forbidden City[N]. The Guardian. 2007-1-18.

XINHUA, 2011. Beijing's tallest skyscraper to be built[N]. China Daily. 2011-9-19.

XIAO X Y, 2011. Crowning glory[N]. China Daily. 2011-10-14.

YANG X F. Management Disaster: A German Ghost Town in the Heart of China[J]. Spiegel on Line. 2011-10-12.

ZHENG J R, 2011. Firm to get skyscraper project off the ground[J]. China Daily. 2011-9-6.

ZHANG L F, 2012. Davos to China: let's try again next year[N]. Financial Times. 2012-1-27.

立于分水岭还是汇入洪流?
——全球化背景下风景园林东方空间观的反思

AT OR IN THE CULTURESHED?
A REFLECTION ON THE EASTERN SPATIAL PROPOSITION OF LANDSCAPE ARCHITECTURE IN THE CONTEXT OF GLOBALIZATION

朱育帆，郭湧，沈洁，张安
博士，教授
清华大学建筑学院

Yufan Zhu, Yong Guo, Jie Shen, An Zhang
Professor, Department of Landscape Architecture
School of Architecture, Tsinghua University

中文翻译：刘佳燕，郭湧
Chinese translation by Jiayan Liu, Yong Guo

在全球化背景下，随着信息与通信技术的发展，人们有机会获取来自国内外各种事件、文化和事物的知识与体验，并能够在全世界范围内共享某些相同的价值观与生活方式。但与此同时，我们也面对全球化带来的以下挑战：全球化是否会导致不同人群价值观与生活方式的单一化？它是否会破坏传统社会及其和谐？它能给不同文明之间的对话带来什么？最后，我们需要什么样的应对策略？

这些问题同样困扰着今天中国的风景园林设计师。面对这些挑战，一些人试图通过划清界限的方式来保护我们曾经辉煌的历史传统。他们站在文化的分水岭上，尽一切努力来避免被全球化的洪流所淹没。

本文提倡的是另一种不同的观点。我们认为应该接受全球化趋势是不可避免的事实，不同文明之间的对话与融合是无法回避的。文中所提及的案例探索了不同的应对方法，但它们都将文化特质以及文化传统中的非物质性内涵作为最关键的内容。它们努力让项目本身与众不同，同时致力于让项目背后的文化为人所知。因而最终非但不会被全球化的洪流所淹没，而且还会通过自身突出的特点对设计界产生影响。在此意义上，它们融入不同文化的"汇水区"中，并且不断地充实着全球化这股洪流。

8.1 沙龙与文人雅集：一种现象

基于不同文脉背景的文化呈现出差异化的物质性特点和非物质性内涵。尽管通常难以用言词来描述这些差异，但却可以通过对某种文化现象进行观察和比较，从而捕捉到一些大致的印象，例如对西方文化背景下的沙龙和来自中国传统文化的文人雅集这两种现象的比较。

In the context of globalization, with the advanced information and communication technology, we have greater knowledge and experience of different events, cultures, and things abroad and also at home. People have been able to share common values and life styles all over the world. At the same time, globalization raises many questions:

Will it homogenize the values and lifestyles of different people in the world?

Is it going to destroy traditional society or its harmony?

What does it mean in terms of the dialogue among civilizations?

And finally, what sort of solutions do we need?

These questions also haunt Chinese landscape architects today. Some of us tried to draw a boundary to preserve our own brilliant historic tradition. They are at the watershed and making every effort to avoid being merged by the flood of globalization.

This article advocates a different proposition: it should be accepted that the trend of globalization is inevitable and there is no way to resist the dialogue and blend among civilizations completely. The cases included in the article see the cultural identity and intangible quality of cultural tradition as essential. They are making efforts to make themselves as well as the culture behind them distinguished. So, eventually, they are rather immerged in the flood of globalization, than influencing the world of design with their outstanding characteristics. In this sense, they are in the watershed, being part of the flood of globalization, and contributing to it.

8.1 Salon and Elites Elegant Assembly: a Phenomenon

Cultures from different contexts present different tangible qualities as well as different intangible characteristics. Although it is very difficult to describe such difference verbally, it is possible to catch some general ideas by observing and comparing certain cultural phenomena, for example, Salon from western cultural context and Elites Elegant Assembly from traditional Chinese cultural context.

Both of the events are regarded as significant events of civilization. Salon is often described as "hostess' living room". It is a gathering of people under the roof of an inspiring host,

这两种聚会都被认为是各自文明中具有重要意义的文化事件。沙龙一般被描述为"女主人家的客厅"，指的是在一位具有启迪思想的主人家中所进行的聚会，人们通过对话一边共同娱乐，一边提升品味，增长知识。

相对而言，文人雅集则与风景、天气以及自然环境联系得更为紧密。诗人与画家们在河畔、树荫下，或是壮观的山水景观之中交流思想，创造艺术。即使有时无法在自然风景中集会，那些当时的精英，中国的文人墨客们也会寻求在具有自然意境的场所中进行交流，例如中国的古典园林。集会地点的重要性甚至能够直接影响诗画作品的创作。

在中国历史上曾经有一个著名的诗会，被称为"兰亭雅集"（公元353年）。文人们在农历三月初三，即春季的修禊日聚集在浙江绍兴的兰亭饮酒作诗。他们约定了一种特殊的饮酒方式：大家沿溪而坐，让酒杯沿着蜿蜒流动的河水顺流而下，当酒杯停在某处时，距离最近的人必须将杯中酒一饮而尽并赋诗一首。这种活动在中国广为流传，被称为"流杯"或是"曲水流觞"。最后，26名参与者一共创作诗词37首，其中还有一部书法作品成为中国文化史中流芳百世的传奇。很难想象如果这些文人不是置身于山川河流的自然风景之中，而只是居于室内彼此交谈，如何可能诞生出这些艺术作品。

通过对比沙龙以及文人雅集，可以明显看出，不同文化特质的形成既离不开物质性的特点，也要有非物质性的内涵。

上述文化现象展现了中国传统文化特色的一个侧面：人们总是与自然保持着紧密的联系。我们可以将其概括为天人合一，人与天调。就空间概念而言，中国传统文化与场所精

held partly to amuse one another and partly to refine the taste and increase the knowledge of the participants through conversation.

Comparably, Elites Elegant Assembly is related much more to the landscape, weather and natural environment. The poets and painters created arts and exchanged thoughts by the river, under canopies of great trees, and with the view of splendid mountains and water. Even sometimes it is difficult to assemble among natural landscape, the Chinese scholars who were concerned as elites of the time, tended to find a site that could inspire the imagination and metaphor of the nature, for example, Chinese gardens. The location of the assembly was so important that the poems and paintings may not able to be composed without proper sites.

In history, there exited a famous poetic event, called Orchid Pavilion Gathering (A.D. 353). Then, literati gathered at the Orchid Pavilion near Shaoxing, Zhejiang, during the Spring Purification Festival, on the third day of the third month of the year, to compose poems and enjoy the wine. The gentlemen had engaged in a drinking contest: wine cups were floated in a small winding creek as the men sat along its banks; whenever a cup stopped, the man closest to the cup was required to empty it and write a poem. This activity was known as "floating goblets", or "Liu Shang" in Chinese. In the end, twenty-six of the participants composed thirty-seven poems. And a piece of calligraphy work was handed down to generations and became a legend of Chinese culture. It is difficult to imagine how these art works came into being, if the gentlemen were not sitting in the landscape, by the creek with floating water but under a roof only through conversation to each other.

From the comparison of Salon and Elites Elegant Assembly, people can find it obviously that the tangible qualities and intangible characteristics shaped the identities of different cultures.

Such phenomena present one aspect of the characteristics of traditional Chinese culture: people always have strong connection with nature. We can conclude it as the harmonization of object and ego and the harmonization of nature and human. In terms of spatial concept, Chinese traditional culture always connects to genius loci and tends to give allegoric meanings to

神紧密联系，并且总是趋向于赋予环境及空间以意境。

8.2 全球化视野背景下的相似特征：一组比较

当我们以全球化的视野进行观察时，可以发现不同的文化背景中也存在着相似的特点。在当今强大的全球化浪潮下，通过比较那些根植于不同文化背景的相似性特点，我们既能增进对自身文化的认知，也有助于对非物质性文化特点形成更加清晰的理解。

8.2.1 比较的背景：全球化

全球化是随着世界观、产品和思想等方面的交流而产生的国际一体化进程。简而言之，全球化指的是全世界范围内自然资源与文化资源的交换过程。交通运输与互联网等通信设施的发达是全球化发展的主要动因，它们促进了经济与文化活动的相互依存。

1980年代中期至1990年代中期，全球化一词得到日益广泛的应用。2000年，国际货币基金组织定义了全球化的4个基本方面：贸易与交易、资本与投资、移民与人口流动，以及知识的扩散。

虽然有学者将全球化的起始定义在现代，但是还有很多学者将其上溯到欧洲大航海和发现新大陆的时代。另有学者甚至将源头追溯到公元前3000年。进入21世纪，全球化进程急剧加速。如果说近年来的全球化趋势引发日益广泛的社会忧虑，主要还是因为这一波强劲的全球化浪潮对不同文化的非物质性特质造成了严重威胁。

可见，全球化并非遥不可及的概念，而是一个正发生在我们身边的过程。无论从物质性还是非物质性层面，它

the environment and space.

8.2 Similar Features in the Context of Global Vision: Comparisons

When observing with a global vision, it is not difficult to find similar features from different cultural backgrounds. It is meaningful to compare with these similar features rooted in different cultures to enhance a clearer understanding of our own cultural identity and intangible characteristics, especially in the context of today's strong trend of globalization.

8.2.1 Context of Comparisons: Globalization

Globalization is the process of international integration arising from the interchange of worldviews, products, ideas and other aspects. Put in simple terms, globalization refers to the processes that promote worldwide exchanges of national and cultural resources. Advances in transportation and telecommunications infrastructure, including the rise of the Internet, are major factors in globalization, generating further interdependence of economic and cultural activities.

The term globalization has been in increasing use since the mid-1980s and especially since the mid-1990s. In 2000, the International Monetary Fund(IMF) identified 4 basic aspects of globalization: trade and transactions, capital and investment movements, migration and movement of people, and the dissemination of knowledge.

Though several scholars place the origins of globalization in modern times, others trace its history long before the European age of discovery and voyages to the New World. Some even trace the origins to the third millennium BCE. Since the beginning of the 21th century, the pace of globalization has intensified at a rapid rate. It is the recent trend of globalization that caused more and more people' s trepidation. Because strong trend of globalization has formed great threaten to the intangible identities of different cultures.

Therefore, globalization is much more than an out-of-reaching concept, but a touchable proceeding phenomenon affecting tangibly and intangibly on the development of human society. It is the trend of globalization that contextualized

都在影响着人类社会的发展。由此，全球化形塑了我们进行比较研究的背景。正是中国独特的文化特质，例如对场所精神、内涵隐喻、情感体验等的关注，成为我们开展比较研究的核心。

8.2.2 比较1：阈

如前文所述，中国传统文化与场所精神紧密相关。对场所精神的阐释与映射是风景园林师在设计工作中所做的主要努力之一。在场地的保护与改造之间实现微妙的平衡是一项很大的挑战。所以在对场所精神进行阐释和传达的时候有必要借助"阈"的概念。

在生理学与心理学领域，"阈"指的是一种界限，当刺激低于该界限时，便无法产生任何反应。当风景园林师利用空间要素重塑场地时，就会对参观者的感知产生某种刺激。当这种刺激足够强烈时，便会形成场地精神与意义的联想。而"阈"便是触发参观者精神体验以及形成物质空间与非物质特征之间意象关联时，设计对场地的最小介入。

对设计师而言，探寻阈值的过程正是理解场地非物质内涵及其与物质条件之间关系的过程。最小介入体现为一种强有力的策略，展现出自然环境与空间设计要素之间的协调关系。

类似于中国传统文化中对场所精神的表达，这一特点也可以在西方当代设计作品中看到。例如由达尼·卡拉万设计的位于西班牙波尔特沃的华尔特·本雅明纪念园（图1）。

波尔特沃镇议会在邀请卡拉万来设计本雅明纪念园之后为他提供了全面的支持，给了他选择场地和材料的充分自主权，以及对现状场地和望景台进行改造甚至拆除的

our comparisons. It is our distinguishing cultural characters concerning genius loci, allegoric meanings, spiritual experience etc. that serve as spotlight, helping us finding the objects to compare.

8.2.2 Comparison 1: Threshold

As previously mentioned, Chinese traditional culture relates to genius loci. The representation and projection of the spirits of the site is one of the main efforts landscape architects make in design works. The delicate balance between the preservation and renovation of the site is a great challenge. It is essential to play with the threshold when try to represent and project the genius loci.

In the terms of physiology and psychology the word "threshold" refers to a limit below which a stimulus causes no reaction. When landscape architects reshape a site with spatial elements, certain stimulus is set to visitors' perception. When such stimulus is strong enough, it becomes redolent of the meanings and spirits of the site. The threshold is the minimum intervention to the site of the design to trigger visitors' spiritual experience as well as the connection between the physical environment and intangible characteristics.

For the designer, the process to approaching the threshold is the process to understand the intangible quality of the site and its relationship to the physical conditions. The minimum intervention is a powerful strategy to reveal the coherence of the natural environment and the designed spatial elements.

The features that similar to such principles rooted in Chinese traditional culture to represent the genius loci can be also concluded from western contemporary design works, for instance, the memorial to Walter Benjamin, by Dani Karavan, in Portbou, Spain (Fig. 1).

Portbou Town Council offered Karavan its wholehearted assistance, giving him complete freedom to choose the site and the materials, and also the freedom to change or even destroy the existing terrace and belvedere, after inviting him to design the memorial to Benjamin. However, the freedom offered to Karavan turned out to be a minimum intervention to the site (Fig. 2).

The original cemetery in the town was kept untouched. The stone marking Benjamin's tomb was preserved as a sign of the

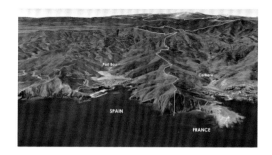

图1 本雅明纪念园位置示意图
（资料来源：朱育帆，改绘自谷歌地球地图）

Fig.1 Location of the memorial to Walter Benjamin
(Source: Yufan Zhu, based on Google Earth map)

自由。然而，卡拉万却巧妙地将这些自由转变成为对场地的最小介入（图2）。

设计原封不动地保留了镇上原有的墓园。作为本雅明墓碑的块石被完整地保留了下来，作为生命逝去的印迹。悬崖边缘迎风伫立着一棵年迈瘦弱的橄榄树，设计师将它尊为为生存而奋斗的象征（图3）。墓园水泥围墙上的铁丝网阻断了望向海平面的视线，设计师将其理解为一种隐喻，比喻无法实现的自由（图4）。现场没有进行任何拆除，相反很多现状都得到了尊重和重新发现。用卡拉万的话说："设计已经存在，我们无法做得更好。我只需要让人们看到它。"

他最终选取了与本雅明墓相邻的一块场地作为纪念园选址。场地面向西班牙与法国的边界，本身已然在讲述本雅明生命的悲剧，现场的海景更是形成了对哲学家一生的隐喻。卡拉万亲手打造了通往严冬风雨之海并指向咆哮激流的通道，为方案赋予了空间形态。橄榄树、块石、路径、海面、海平线、围墙、墓园都饱含寓意。在卡拉万看来，这位哲学家的命运可以通过风景自身表现的环境现象来阐释，他的工作便是通过对现状环境的最小介入让人们感受到这种阐释（Restany, 1992）。

在设计中谨慎地接近阈值以探索对场地的最小必要干预，进而引入尽量轻质的结构以表现场所精神，这种设计策略也曾在我们的设计项目中加以应用，那就是上海辰山国际植物园的矿坑花园。

大约300年前，该项目所在区域曾经是城市外围著名的风景地。辰山周围的景致在地方志中被骄傲地记载为"辰山八景"。近一个世纪以来，场地被用作采石场，为上海的城市发展提供建筑材料。2007年，这处业已废弃

lost life. The old little olive tree standing on the edge of the cliff against the wind was also respected as a symbol of the fighting for surviving (Fig. 3). The wiremesh on the cemetery wall blocking the view to the horizon was understood as a metaphor of the broken of the promised freedom (Fig. 4). Nothing on the site was destroyed and much was respected and re-discovered. In Karavan's word: "the work is already there, I can't do better. I only have to make people see that."

He finally chose the memorial a place adjacent to the tomb of Benjamin, which facing the boundary between Spain and France. The site was already telling the tragedy of Benjamin's life. And the phenomenon of water served as a metaphor for the life of the philosopher. Karavan framed with his hands the path of stormy winter sea containing the vortex and gave form to his project. Olive tree, stones, path, sea, horizon, fence, cemetery were all meaningful. Karavan saw that the philosopher's fate could be read in the elemental signs with which the landscape expressed itself. His work would be to make it perceived with minimum intervention to the existing environment.

Approaching to the threshold in design, detecting the minimum necessary intervention to the site, and introducing as light structure as possible to represent the spirit of the site, such design strategies were also implemented in one of our design projects: Quarry Garden in National Botanical Park, Chenshan, Shanghai, China.

About 300 years ago, the area was one of the famous scenery landscapes outside Shanghai. There were eight scenes around the Mountain Chenshan, which were recorded in the local chronicles proudly. In the recent century, the site was used as quarry site to offer construction materials to the development of the metropolitan area. In 2007, the out-of-used quarry site was planned as a part of new National Botanical Park in Chenshan, Shanghai.

For most of the people the site was a scar on the land (Fig. 5), but as a landscape architect that was trained as a Chinese traditional gardener, the designer visited the site and captured the first image of it differently. For the designer, some expressions of the site were like scenery paintings. Although the site functioned as a quarry yard for decades and was left behind as a pit barely accessible, the strong sense of closure

图2 通往严冬风雨之海的通道
（资料来源：朱育帆，改绘自谷歌地球地图）

Fig.2 Path of stormy winter sea
(Source: Yufan Zhu, based on Google Earth map)

图3 引入数级钢质台阶将现状橄榄树转化为设计的组成部分
（资料来源：朱育帆）

Fig.3 Several steel stairs introduced to converse existing olive trees as a part of the design
(Source: Yufan Zhu)

的采石场被重新规划，作为新建的上海辰山国家植物园的一部分。

对于大多数人而言，这个场地无非是大地的疤痕（图5）。但是作为一个具有中国传统园林设计教育背景的风景园林师，设计师在踏勘场地后形成了完全不同的第一印象。对于设计师而言，场地所展现出的一些状态暗合山水画的表达。数十年间作为采石场的经历，在这里留下了一个几乎不可进入的深坑，而其中强烈的围合感以及崖壁上令人惊叹的纹理让设计师瞬间联想到传统中国山水画中的笔墨皴法之美（图6）。一个念想涌入设计师的脑海，即"设计已然存在"。

那么设计问题就转化为如何让参观者对这种感知产生共鸣。如何让场地中既有的美能够被没有风景园林师专业眼力的参观者感受到呢？作为回应，有必要引入一套空间要素，允许参观者们深入到场地中曾经难以达到的位置，深度触碰场地，并让他们理解物质空间背后的非物质特性。

在中国文化中，人们从风景中感知到的非物质性内涵与人们在环境中的行为之间通常有着紧密联系。桃花源就是一个经典案例，它反映了中国人脑海中的理想社会模型，正如伊甸园对于西方的意义。下面的文字描述了桃花源被发现的过程，及其独特的神秘入口。

"晋太元中，武陵人捕鱼为业。缘溪行，忘路之远近。忽逢桃花林，夹岸数百步，中无杂树，芳草鲜美，落英缤纷。渔人甚异之。复前行，欲穷其林。林尽水源，便得一山，山有小口，仿佛若有光。便舍船，从口入。初极狭，才通人。复行数十步，豁然开朗……"

基于共同的文化背景，中国人往往会有一种类似的期

and the marvelous texture on its cliff reminded him the beauty from traditional drawing technique of Chinese scenery paintings immediately (Fig. 6). One idea ran into the mind that "the design is already there".

Then the design question turned into how to let the visitors have the echo of such perception. How to let the beauty existing on the site perceived by the visitors even without professional eyes that landscape architects have. To respond to such challenge, it is necessary to introduce a set of spatial elements to allow the visitors to touch the site from an impossible spot that deep inside the place. And let them understand the intangible quality behind the physical space.

In Chinese culture, the intangible quality that people perceived from landscape is strongly related to the way people act in the environment. The Peach-Blossom-Grove is such an example, which reflects an ideal model of a utopia for Chinese, as the Eden to the westerners. The following text described how the place was discovered and its particularly mysterious entrance.

"In the Period of Taiyuan of the Jin Dynasty, there was a fisherman whose native land was called Wuling.Once a time, he was walking along a creek-riverside, with no sense of how far away from his start. All of a sudden, he was surprised to find a Peach-Blossom-Grove, and on both sides of the creek hundreds of feet long with no other trees but peach ones grow with fresh grass and fallen flowers. Being surprised and going on forward, he passed through the whole grove, reaching a fountainhead near a mountain, with a small hole seemingly glittering. He left his boat; entering the hole which was so narrow that only one man could press himself to penetrate into it. After dozens of steps walking, he found the view instantly clearing up".

With such common cultural background, Chinese people share a sense of expectation to something surprisingly beautiful when mysterious entrance is set ahead. So, theoretically, it is possible to trigger visitors' perception of the existing beauty with nothing more than a path starting mysteriously and a scene that evokes people's imagination and satisfies people's curiosity.

This is why on the quarry site, the completed project was composed with only an enclosed box leaning ahead to the pit, pouring people forward (Fig. 7). A path carefully stretching

图4 望向海平线的小平台，混凝土墙上的铁丝网象征哲学家无法实现的自由
（资料来源：朱育帆）

Fig.4 Small platform facing the horizon over the wiremesh on the cemetery wall representing the broken of promised freedom of the philosopher
(Source: Yufan Zhu)

图5 上海辰山矿坑改造前状态
（资料来源：朱育帆）

Fig.5 Existing situation of the quarry pit in Chenshan, Shanghai
(Source: Yufan Zhu)

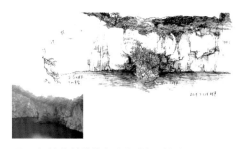

图6 场地的某些状态让人联想到山水画
（资料来源：朱育帆）

Fig.6 Situation of the site evoking the sense of scenery painting
(Source: Yufan Zhu)

图7 封闭的盒子在坑壁边缘向前倾斜，将人们向坑内"倾倒"
（资料来源：陈尧）

Fig.7 Enclosed box leaning ahead to the pit, pouring people forward
(Source: Yao Chen)

图8 沿着坑壁延伸的路径，尽可能地减小对崖壁造成的破坏
（资料来源：陈尧）

Fig.8 Path stretching along the cliff with the minimum impact to it
(Source: Yao Chen)

待心理，即，当面对某个神秘狭小的入口时，会期待"豁然开朗"的惊艳与美丽。因此，从理论上而言，欲激发参观者对现场存在之美的感知，只需要提供带有神秘感的路径和能够引起人们想象共鸣与满足好奇的"豁然开朗"的场景。

这就是为什么最终矿坑方案的构成要素非常简单：只有一个封闭的盒子，在坑壁边缘向前倾斜，将人们向坑内"倾倒"（图7）；一条沿着崖壁小心展开的路径，尽可能地减小对崖壁造成的破坏（图8）；一个悬挂在路径对面的瀑布，无论谁下到坑底都不会错过（图9）。

上述两个项目都采取了类似的设计原则和某些相似的设计语言。它们都尊重现状环境，采取最小的介入来构建物质环境与精神体验之间的联系，从而展现场地的非物质性特质。但每个项目又有独特的特点。

本雅明纪念园将环境作为隐喻。卡拉万成功地让现状中的自然环境和建成环境来讲述哲学家的生命。他引入的空间元素很轻，但却构建起人的情感与风景之间的强烈联系。这种联系进而放大了人们对自然现象的感知，反过来又强化了参观者的情感。在方案中，是环境将本雅明的精神具体化了。在上海辰山矿坑花园项目中，重塑场地的不仅是具体的建造，更重要的是深深存在于人们意识中的共同文化意象。非物质性文化意象的映射重构了场地的概念。所以，非物质性的传统文化因素与所有物质材料一起共同构成了设计的关键性内容。

8.2.3 比较2：互惠

建筑与风景之间的互惠是中国建筑与园林传统文化中的一个核心特点。中国文人精神崇尚意境与人化自然，对

along the cliff with as little impact to it as possible (Fig. 8), and a water cluster hanging opposite to the path, that impossible to miss for whoever gets down to the pit (Fig. 9).

The two projects share similar design principles as well as certain design languages. Both projects respect the existing environment and took minimum intervention to build connection between physical environment and spiritual experience to represent the intangible identity of the site. But each project shows unique characteristics.

The memorial site takes the environment as a metaphor. Karavan managed to let the existing natural and cultural environment tell stories of the philosopher's life. The spatial elements introduced on site are light, but establish strong connections between people's feelings and the scene. Such connections amplify people's perception to the power of the natural phenomenon and in return enhance the feelings of visitors. The environment materializes the spirits of Benjamin. In the project of quarry garden in Chenshan, Shanghai, it is the common cultural image deeply embodied in people's conscious, rather than the specific construction that converses the site. The projection of the cultural idea conceptualizes the site. So, the traditional culture is equally important to any physical material, and is the essential part of the design.

8.2.3 Comparison 2: Reciprocity

The reciprocity between architecture and landscape is a genuine characteristic in Chinese traditional culture. Chinese scholar spirits, which appreciate the allegoric meanings and the humanized nature, contribute greatly to such inherent value of Chinese cultural tradition. Taking the Elites Elegant Assembly as an example, cultural activities happening with the assembly had strong connections with natural environment. As well as the paintings, poems and calligraphy, the related architectures were also connected closely to the nature. Such cultural phenomenon was representatively expressed in Chinese traditional gardens. One of the design principles of Chinese classical gardens is to "borrow" scenes from surrounding environment. In other words, it is the principle of connecting the garden to a vast landscape visually and conceptually. With such design principle, architecture and landscape are operated

这种传统文化的内在价值贡献很大。例如在雅集中，集会过程中的文化活动与自然环境具有强烈的联系。与绘画、诗赋、书法一样，相关的建筑也与自然紧密相连。这种文化现象突出地反映在中国传统园林中。中国传统园林的一条设计原则就是从周围环境中"借景"。换言之，就是从概念上和视觉上将园林与更广阔的风景相联系。在此设计原则下，建筑与风景的营造之间体现出互惠性。正是这种互惠性的营造，消除了两者间在物质上和概念上的相互分离。

意大利威尼斯的奎里尼·斯坦帕里亚基金会因为卡罗·斯卡帕的成功改造而声名远扬。英国建筑师、评论家肯尼特·弗拉姆顿从中看到了中国传统园林对斯卡帕作品的影响，并且他相信这种影响非常深刻，再怎么强调也不为过。在这个项目中正体现出建筑与风景彼此互惠的设计。

斯卡帕将建筑的改造与潮水涨落的具体现实之间建立起结构性关联。潮水涨落这一不可避免的城市背景为设计带来了最关键的挑战。斯卡帕的应对策略是对建筑进行重建并改造庭院的布局，来重新梳理奎里尼别墅的内部交通流线和建筑首层的使用功能。室外的运河被引入别墅的内部（图10）。通过在室内增建水位控制系统形成的景观从概念上将室内空间室外化。19世纪以后的建筑加建部分在改造中被拆除，使内厅回复到最初的结构。建筑入口在改造中被移到建筑前面连接洪泛防护系统的一个小广场上（图11）。

改造之后，水变成了设计的核心材料。建筑的内部空间由于水的动态过程而得以重塑。根据潮水水位的变化，房间内既可以完全地排出积水，也可以全部淹没于水中。

with reciprocity. The operation of reciprocity diminishes physical and conceptual separation between both.

Querini Stampalia Foundation, in Venice, Italy is a famous project with its successful modification by Carlo Scarpa. The British architect, critic, Kenneth Frampton saw great influence of Chinese traditional gardens to Scarpa's work, and he believed that such influence could hardly be overestimated in the project of Querini Stampalia, which is a project of the operation of reciprocity between architecture and the landscape.

Scarpa's modification to the palazzo is dependent on structural relations to the specific reality of the rising of the tides. This unavoidable urban context brought the main challenge. Scarpa's strategy to respond was to restore the portego and the layout of the garden to restructure the circulation inside the Palazzo Querini, and change the use of the first floor. The canal was engaged within the interior of the palazzo. The interior was conceptualized as an exterior, by creating a landscape of water control structures within it (Fig. 10). The extraneous nineteenth-century additions was removed to recover the original space of the portego, and the entrance was moved to a small square connecting the high water protection system, in front of the palazzo (Fig. 11).

After the modification, water has become the genuine materials of the design. The space of the portego was reshaped with the dynamic process of the water. The room can be completely free of water or completely immersed, according to the different level of the tide. The interior landscape operates as an apparatus that measures natural phenomena, revealing the unpredictable and dynamic characteristic of the exterior environment. The garden was an extension of the space. Compared to the dynamic process in the portego, the controlled and guided water was the main theme of the garden. Within the raised lawn, guided by a channel water flowed from a small alabaster labyrinth-pond, past a stone lion and through a duct to disappear in a curve-shaped well. As the natural process was introduced into the portego, the conceptual characteristic was introduced into the garden. The controlled and guided water was an expression of the culture of symbiosis with water in the city of Venice.

Scarpa made the Palazzo Querini a physical carrier of the

图9 路径对面悬挂的飞瀑
（资料来源：陈尧）

Fig.9 Water cluster hanging opposite to the path
(Source: Yao Chen)

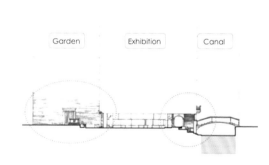

图10 奎里尼·斯坦帕里亚基金会建筑剖面图
（资料来源：沈洁，改绘自 Anita Berrizbeitia, Linda Pollak）

Fig.10 Section of the Querini Stampalia Foundation
(Source: Jie Shen, based on Anita Berrizbeitia, Linda Pollak)

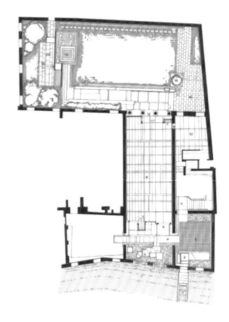

图11 奎里尼·斯坦帕里亚基金会建筑平面图
（资料来源：沈洁，改绘自Kenneth Frampton，王骏阳）

Fig.11 Plan of the Querini-Stampalia Foundation
(Source: Jie Shen, based on Kenneth Frampton, Junyang Wang)

图12 金融街小院周边环境
（资料来源：朱育帆）

Fig.12 Surrounding circumstance of the courtyard
(Source: Yufan Zhu)

室内景观成为测量自然水位的标尺，展现出外部环境不可预测的动态特征。与建筑内部空间的动态过程相比，建筑外部庭院的主题则是围绕受控的、受引导的水体展开。在抬起的草地上，水流在水渠的引导下从一个小的大理石迷宫水池中流出，经过石狮，通过管槽，最终跌入一处精致的曲线型的井中。当自然水文过程被引入建筑中庭，意象化的特征亦被引入庭院。管槽中受控的、渠化的水体正是威尼斯水与城共生的文化表达。

斯卡帕通过将建筑概念化为一种动态的自然过程，用微妙的细节营造庭院，并用隐喻的手法映射城市尺度上的运河建造，从而将奎里尼别墅营造成为非物质性内涵的物质载体。这种设计通过室内与室外空间的整合与延展消除了建筑与外部环境之间的分隔，同时提升了两者的品质，由此达到建筑与风景之间的互惠。

下一个比较案例是北京金融街13号院的改造项目，以此解释当代中国设计手法中对于"互惠"的阐释。这个院落过去是老北京城中一座道观的组成部分。当前周边区域已经开发成首都的金融区，几乎没有能够反映历史信息的背景得以保存。院落虽然由于具有文化价值得到保护，但是现在孤立地座落在机动车道环绕中的一处绿化带中间（图12）。在重建之前，这座院落的保护状况很差，由于缺乏必要的日常维护，建筑已经成为危房。院落空间也被随意搭建的部分蚕食，变成了杂院。

项目的目的是创建一个文化沙龙，可以让参观者脱离城市的商务环境，在轻松的诗意氛围下聚会。为了创造能够让来访者联想到文化体验的非物质性内涵，这个院落被作为场所精神的载体来对待。建筑与庭院在整个设计过程中从来没有分离。互惠性的营造是设计的核心概念。自晋代以来

intangible quality by meanings of conceptualizing the building as a dynamic natural process, constructing the garden with subtle details as well as with the metaphor of the construction of the canals of the city in macro scale. Such design diminished the separation of the building and the garden with the integration and extension of the interior and exterior space and improved the quality of the both. The reciprocity between the building and the garden thus can be achieved.

The project we would present as a comparison to try to explain the contemporary Chinese interpretation of reciprocity is the rebuilding of No. 13 courtyard in Beijing Financial District, China. The courtyard used to be a part of a Taoist temple sitting in the old Beijing city. Now, the surrounding area has been developed as the financial district of the capital and had little context reflecting any historic information left. The courtyard was preserved as a cultural heritage and located in a green island isolated by the vehicle traffic (Fig. 12). Before rebuilding, the courtyard was badly preserved. The buildings were in dangerous condition due to the lack of necessary maintenance. The space of the courtyard was crammed by hastily constructed extraneous additions.

The objective of the project was to create a cultural salon allowing visitors meeting in a relaxed poetic atmosphere rather than stressful urban business atmosphere. In order to create the intangible quality connecting the visitors to cultural experience, the courtyard was treated as a main carrier of the place spirits. The buildings and the yard were never separated in the whole process of designing. The operation of reciprocity is a genuine idea of the design. And the Chinese scholar spirits from Jin Dynasty, which has deeply influenced the expression of Chinese gardens, was interpreted in the design. In order to interpret such scholar spirits, ideas from Chinese classical gardens were borrowed.

The rebuilding started with the removing of the additional parts that damaged the quality of the buildings and the yard. When the extraneous additions were removed, the original materials and structure of the buildings were revealed. And the interfaces of the courtyard were accordingly rediscovered. Then new structures were introduced not only for enhancing the stability of the old buildings, but also for reshaping the space

深刻影响中国传统园林面貌的中国文人精神在设计中得到阐释。为此，项目借用了中国传统园林中的一些意象。

院落的重建从拆除那些破坏建筑和院落品质的加建部分开始。当加建部分拆除后，建筑原初的材料和结构就显现出来，而且院落的界面也相应地恢复了。然后引入新的结构，不仅增强了老建筑的稳定性，也重新塑造了空间，将建筑与院落之间的隔离消解。方案在院落入口的位置加建了一片新的墙体作为影壁，使用的是传统的材料与处理工艺，但是在砌筑方法上做了创新。墙体的引导整合了动线，可以提供室内与室外结合的空间体验。它还形成了一种新与旧、记忆与感知的对话，开启了物质环境意象化的序列。

项目的核心空间是庭院。在庭院中，具有象征意义的是横卧于碎石地面的石块。在强烈的体量对比下，石块成为整个空间的焦点。它看上去就像轻浮在地表上，展现出一种好客的友好姿态。在庭院的一角，一个白色的浅圆盘放置在碎石铺装上，好像是悄悄从地下浮出来一样。在院落的边缘设计了一个小型水景，上面是如镜般的水面，侧方有锈钢板包裹支撑，溢出的水流顺着锈钢板缓缓流淌。在这个水景背后，一丛纤竹倚壁而立。当微风徐来，纤细的竹叶轻摇细语。院中唯一的乔木是那棵古枣树。它自始至终伫立着，见证了场地的记忆（图13）。

庭院中所有的要素共同构成了一个关于文人雅集的喻义。从地面浮起的白色圆盘是水中月影的象征。在中国传统文化中，月亮长久以来在无数的文学作品中被拟人化。她是一个通情达理的伙伴，孤单时可以对其倾诉。竹子代表清风，通过动态和声音将风的姿态视觉化。在传统文化中竹子和清风都是高尚道德与崇高精神的象征。石，则

and diminishing the separation between the buildings and the courtyard. A new wall constructed with traditional material and its technical treatments however with the innovated construction measures was placed at the doorway as a threshold. It created an initialized circulation that offered integrated interior and exterior spatial experience. And it also generated a sense of dialogue between the old and new, the memory and perception. It started a sequence that conceptualized the physical environment.

The core space of the project was the yard. In the yard, the symbolic feature was the stone horizontally laid on the surface of gravel. With the strong contrast of the quality and volume, the stone formed the focus of the entire space. It looked like floating on the surface lightly with a gesture of hospitality. At the corner of the yard a shallow white basin was planted in the gravel surface as if it was emerging quietly to the surface to join the meeting. At the edge of the yard, a small waterscape was designed with a mirror face of water and slight effusion over rust-steeled supporting panels. A clump of bamboo was planted against the wall at the back of the waterscape. As wind blew, the fine leaves of the bamboo shook and whispered. The only tree in the courtyard was an old date tree. It had always been there as the memory of the site (Fig. 13).

All the elements in the courtyard composed a metaphor of Elites Elegant Assembly. The white basin emerging from the surface is a representative of the reflection of the Moon in the water. In Chinese traditional culture, the moon has long been humanized in countless literature. She is an understanding companion of the litterateurs with whom one can talk to when feeling lonely. The bamboo is the representative of fresh wind. The movement and sound of bamboos visualize the movement of wind. Both bamboo and fresh wind are symbols of honorable moral and respective spirits in the traditional culture. And the stone is the materialized spirits of the owner of a garden or the author of a literature. In other word, it is an expression of "me". Such metaphor is original from an ancient poem by Su Shi, a Chinese famous poet in the Song Dynasty. The poem reads as the following:

Leaning against the bed in leisure and looking out to the mountains.

图13 金融街小院中心庭院
（资料来源：朱育帆）

Fig.13 Center of the courtyard
(Source: Yufan Zhu)

是园林主人或者文学作者的精神物化，也就是"我"的象征。这种比喻诠释意境源自中国著名诗人苏东坡在《点绛唇》中的诗句：

闲倚胡床，庾公楼外峰千朵。与谁同坐。明月清风我。

由于庭院的设计，建筑内部的活动随之意象化。集会成了与高朋之会，与"我"同道之会。因此，建筑与风景之间的互惠不仅停留在空间和感知层面，更深入到非物质性的精神层面（图14）。

上述两个项目虽然根源于不同的文化，面对的条件也各异，但是对空间的理解和设计的原则却是相通的。它们都将室内和室外空间进行了整合。虽然没有直接的思想上的交流，但是二者都在全球化的总体进程中深受中国传统园林的影响。

使二者彼此区别的是非物质性的内涵以及文化传统。在奎里尼·斯坦帕里亚基金会项目中，斯卡帕根据对水文现象的精辟理解，将建筑向城市文脉开放，把室内建筑空间与潮汐的自然过程结合，并将室外空间的意象赋予室内空间。相对而言，北京金融街13号院并没有强调与周围环境的联系。因为院落的周边环境已经发生了极大的变化，只能将室内环境与概念上的外部环境相关联。这种关联通过"清风、明月和我"的比喻实现。在中国传统文化背景下，人们并不难理解，这种比喻是关于人化自然和文人雅集文化活动的意象化。所以，如果没有中国传统文化非物质性的内涵，金融街小院改造的设计就无法实现。

8.2.4 比较3：路径

路径决定了空间中的行为与运动过程，也定义了感知的序列。随着空间与时间内涵的组合，路径可以形成叙事

whom should I sit with?
The bright moon, the fresh wind and me.

Poem of Dian Jiang Chun

With the design of the courtyard, the activities inside buildings are conceptualized, which is to meet friends with noble personality and people who share consonance in feelings with "me". Thus, the operation of reciprocity between the building and landscape is not only physical and perceivable but also intangible and spiritual (Fig. 14).

The two projects although rooted in different cultures and faced different conditions, shared common understandings of space and design principle, which is to integrate the interior and exterior space. Although without direct exchange of ideas, both of them were influenced by Chinese classical gardens, in a general process of globalization.

What distinguished each other in the projects were the intangible qualities and cultural traditions. In the project of Querini Stampalia Foundation, with a profound understanding of water phenomena and culture, Scarpa opened the building to the urban context; connected the interior architectural space to the natural process of tide; and conceptualized the interior space as exterior. Comparably, the project of No. 13 courtyard in Beijing Financial District didn't emphasize the connection to the context. Because the courtyard's context has dramatically changed, it was only possible to connect the interior environment of the courtyard to a conceptualized exterior context. The connection was realized with the metaphor of the moon, the fresh wind and the expression of "me". With the background of Chinese traditional culture, it is not difficult to understand that this metaphor is about a humanized nature and conceptualized cultural activities of elegant assembly. So, the design couldn't be considered completed if the intangible quality of Chinese traditional culture was not counted in.

8.2.4 Comparison 3: Passage

A passage determines the process and acts of the movements on site. It also defines the sequences of the perception. With the combination of spatial and temporal qualities, a passage forms a narrative process, which makes it a carrier of intangible characteristics of the design. From the

图14 室内活动意象化
（资料来源：左图：孙位；右图：朱育帆）

Fig.14 Activities inside buildings are conceptualized
(Source: Left: Wei Sun; Right: Yufan Zhu)

性过程，可以成为设计非物质性特质的载体。从东方空间观出发，路径的组织是外部空间设计中最为重要的表达情感和投射情绪的策略。例如，中国传统园林中有"步移景异"的设计原则。遵照这一原则，即使是很小的园林也可以呈现出多样化的景致，并向来访者传达不同的情绪和象征意义。相似的策略也可以在东西方的现代设计中看到，尤其是需要叙事性特点和象征意义的纪念性景观设计。

与其他著名的纪念性景观项目相比，英国杰夫瑞·杰里柯设计的肯尼迪纪念园是很有特色的一个。它由非常简单的空间元素构成，但是却包涵着丰富的象征寓意。其中的路径是赋予项目空间形态与思想寓意的重要构成元素。项目位于兰尼美地草原一块北望泰晤士河的高地上（图15）。一条由6万多块象征朝圣人群的花岗岩石丁铺就的蜿蜒小径一直延伸到肯尼迪总统的纪念碑。这条小径从一处柴扉起始，颠簸上行穿过一片自然生长的树林，直抵纪念碑。据设计者介绍，沿路径的旅行是一次关于生命、死亡与精神的旅程。树林是自然生命力充满活力和神秘的象征。为了渲染一种阴郁的氛围，这片野生树林的边缘部分加植了杜鹃以增加种植厚度。地被层也进行了强化。于是，从林地直到纪念碑，人们在林地里的视野中看不到任何开阔地。纪念碑象征的是被众人肩扛的棺椁。在纪念碑之后的部分，路径转变为规整的形式，引导人们到达用于冥想的隐蔽座位，越过兰尼美地的田野俯瞰泰晤士河。

杰里柯解释道："座位、道路和门限的形式是严格按照比例的几何形，正如将和谐从天国带到人间的希腊神庙。"在这种寓意下，纪念碑之后的道路象征的是永生（Spens, 1994）（图16）。

eastern space concept, the organization of passages is one of the most important strategies to project the feelings and emotion to the participators of a designed exterior space. For example, one of the design principles of Chinese classical gardens is to present a different scene on every different step during the walk along the passage. With this principle, even a very small garden can express great variety of scenes and project different emotions and symbolic meanings to visitors. Similar strategies can also easily be found in modern designs, either eastern or western, especially in memorial landscape designs, in which narrative characteristic and symbolic meanings are highly appreciated.

Among famous memorial landscape projects, the Kennedy Memorial by Sir Geoffrey Jellicoe in Runnymede, England is one that composed with simple spatial elements but strong symbolic meanings. The passage was the singular component that shaped the project and made it allegorical and emotional. The project located on a piece of rising ground looking north over the river Thames, at the meadow of Runnymede (Fig. 15). A winding path made up by 60,000 granite setts represented pilgrims jostling their way upwards to the monumental stone of President Kennedy. The path started with a wicket-gate. It groped upwards through a self-regenerating wood to the monumental stone. According to the designer, the journey along the path is one of life, death and spirit. The wood is a symbol of virility and mystery of nature as a life force. In order to enhance the sense of foreboding, the edge of this piece of self-regenerating wood was thicken with rhododendron edging. The ground cover was also encouraged. So that no open land would be seen from the woodlands until the monumental stone was reached. The stone was a metaphor of a catafalque balanced on the shoulders of the populace. Beyond it the passage changed the form into a formal path, which lead to the secluded seats for contemplation overlooking the fields of Runnymede.

Jellicoe said: "The form of seats, paths and threshold, like a Greek temple that brings the harmony of the heavens to earth is proportioned geometry." With such meaning, the path beyond the monumental stone represented immortality (Fig. 16).

The Kennedy memorial is an adventure of allegory. Instead of shaping a monumental place, the designer created a process,

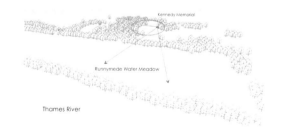

图15 肯尼迪纪念园坐落在向北俯瞰泰晤士河的高地上
（资料来源：沈洁）

Fig.15 Kennedy Memorial sitting on the rising ground looking north over the river Thames
(Source: Jie Shen)

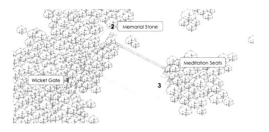

图16 路径是给设计赋予空间形态的核心要素
（资料来源：沈洁）

Fig.16 Passage is the singular component that shaped the project (Source: Jie Shen)

图17 原子弹纪念公园周边情况
（资料来源：朱育帆工作室）

Fig.17 Context of the unclear memorial park
(Source: Yufan Zhu Studio)

图18 场地上茂密的青杨
（资料来源：左图：朱育帆；右图：朱育帆，改绘自谷歌地球地图）

Fig.18 Thriving poplar trees on the site
(Source: Left: Yufan Zhu; Right: Yufan Zhu, based on Google Earth map)

肯尼迪纪念园由此成为一个富有寓意的场所。设计师并没有塑造一个纪念场地，而是创造了一种过程，来访者在其间可以体会设计师的情感，而不是仅仅作为外围的观察者进行观看。虽然这条路径自身在场地中几乎不可见，但是它确实引发了非物质性的体验，并给整个风景带来特殊的氛围。

用来进行比较的案例是青海原子城的国家爱国主义教育基地——原子弹纪念公园（图17）。虽然意想不到，但也不足为奇的是这个项目也采用了与肯尼迪纪念园类似的策略：设计一条路径，在现状空间要素之间形成新的联系，并塑造场地的象征意义。

公园设计的目的是纪念中国核工业的先驱。1960年代成功地独立发展核工业是中国里程碑式的事件。这对于一个历经了一个多世纪积贫积弱、饱受侵略的民族而言，在重塑民族自信方面具有重要意义。公园场地曾经是荒芜的戈壁滩，也是核技术的野外实验场。在这样的场地上建设纪念园，从物质实体和精神层面上与那段特殊的历史建立联系就非常重要。

在对场地进行调查之后，设计者提炼出三个方面的特征要素。最为突出的要素是那些茂盛的杨树，能出现在海拔3100米的场域实在尤为珍贵。在环境极其恶劣的戈壁滩上，杨树的成活几乎不可能。但是这些由先驱们亲手种下的树木不仅成活而且繁茂地生长。它们是先驱们奋斗历程的鲜活记忆，也是他们坚强意志和奉献精神的象征（图18）。

第二类特征要素是表面留有特殊肌理的锈钢板。它们并非直接取自公园所在的场地，而是来自于相距不远的核爆实验场。在实验场上建有厚重的地下掩体，可以从内部

which the visitors were involved in and experienced the emotions of the designer, rather than only watching as outside observers. Although the passage itself was barely visible, it indeed brought intangible experience and the atmosphere to the entire landscape.

The project selected to compare with the above is the landscape design of the nuclear memorial park of national patriotism education bases in Qinghai Atomic City, China (Fig. 17). Unexpectedly but not surprisingly, a similar strategy was applied in this project as that in the project of Kennedy Memorial. A path was designed to build new connections between the existing spatial elements and to create allegory meanings of the site.

The park was planned to memorialize the pioneers of Chinese nuclear industry. The success in developing the nuclear industry independently in the 1960s was regarded as a landmark event of the People's Republic of China. It was of great significance to reshape the confidence of a nation just recovered from century-long weakness, poverty and invasions. The site of the park was atrocious Gobi desert as well as the core site of the field experiments of nuclear technology. To build a memorial on such a site, it was important to find the connections to that special period of history, either physically or conceptually.

After the investigation of the site, three important elements were found to characterize the project. The most impressive elements were thriving poplar trees, which are precious at the altitude over 3,100 meters. Under the atrocious circumstance, it was almost impossible for the poplar trees to survive in Gobi desert. But these trees planted by the pioneers themselves by hand not only survived but also thrived. They are the alive memory of the history of the pioneers and the symbol of their strong will and honorable dedication spirit (Fig. 18).

Another element was the rusted steel with special texture on the surface. They were not directly from the very site of the park but the field experimental site of atomic explosion not far away. On the experimental site, heavy underground bunkers were constructed, from the inside of which the process of the explosion could be observed. After the explosion, the steel surface of the bunkers would be contaminated with radiation. For

观察爆炸过程。爆炸之后，掩体的钢铁表面会受到放射性污染。为了安全，必须探测并清除污染物。将受污钢铁表面进行切割清理之后就会留下凹凸不平的痕迹，最终形成特殊的表面肌理。所以，锈蚀钢板和这种特殊的切割肌理将现有的场地与历史的记忆相联系（图19）。

第三类要素不是场地上的物质性内容，而是先驱们的事迹。其中有一个关于一对年轻科学家夫妇的故事。一对刚完婚不久的青年物理学家，夫妻二人分别接受了保密的核试验研究任务。他们按照要求严守秘密，隐姓埋名，搬离北京入驻戈壁深处的研究基地。两位年轻的科学家严格地遵守着各自的命令，没有将自己任务的任何信息透露给任何人。他们在戈壁滩中工作了多年，妻子以为她的丈夫一直在千里之外的北京等待自己回家，而丈夫也以为自己的妻子等在家中。他们只能通过写信的方式联系彼此。信件先按照信封上的地址寄往北京的家中，然后由政府部门转投到他们的实际所在地。一封信的投递需要花上几个月的时间。在此过程中他们一直为国家严守秘密，从没有透露过自己的真实位置。数年之后，核爆的日子终于到了，他们都出色地完成了各自的任务。在欢庆的人群中，这对夫妻意外地发现了彼此。直到此时，他们才知道，原来两个人一直生活在同一个基地里，二者之间的直线距离从来没有超过200米。

为了将上述要素整合，方案设计了一条路径形成叙事性过程。场地上的杨树是决定性要素。它们曾经作为建筑的附属物加以栽植，遵循建筑的空间秩序。当原有建筑被拆除之后，它们就成为场地新空间秩序的构架。路径的设计非常仔细，一方面要嵌合到新的空间秩序中，另一方面要连接原子城博物馆与和平之丘两个节点（图20～图22）。路径

safety, the contamination had to be detected and removed. After cutting away the contaminated steel, there left many unevenness on the surface, which formed the special texture in the end. So, the rusted steel and such special cutting texture connected the current site to the memory of the history (Fig. 19).

The third element was not tangible quality on site but stories of the pioneers. Among the stories, there was one about a couple of scientists. The just married couple was both physical scientists. The husband and wife were recruited to the highly classified projects of atomic experiments separately. They were required to keep secret, turn anonymous and move from Beijing to the secret base deep inside Gobi desert to join the teams. Both young scientists followed the orders strictly and didn't tell each other any information about their missions. For years they worked in Gobi desert. The wife thought her husband was waiting for her thousands kilometers away in Beijing at home and the husband thought she did. They tried to contact each other by mail. The mail would be delivered firstly to Beijing according to the address on the envelope. Then, the government would redeliver it back to the base where they actually were. Not until months later, could they hear from each other. As always, they never leaked any information of the projects and their real locations. Years after, the day of the atomic explosion came, their missions were successfully completed. In the crowd of the celebration, surprisingly the couple found each other. Not until then, did they know that for years, they had lived in the same base and the distance between them was no farther than 200 meters.

A path was designed to form a narrative process by combining all the above-mentioned elements. The poplar trees on the site were the defining feature. They had been planted as appurtenance of buildings, following the spatial order defined by the buildings. When the old buildings were removed, they turned out to be the structure of new spatial order on the site. The path was designed elaborately on one hand to fit the new spatial order, on the other hand to connect the newly built Museum of Atomic City and the Hill of Peace (Fig. 20–Fig. 22). The two ends of the path created a hidden axis and the hill stood as the focus of the park both conceptually and spatially. Symbolically, the hill represented the peace of the world, which was the utmost goal

图19 实验场上的掩体及其表面的特殊肌理
（资料来源：朱育帆）

Fig.19 Bunker on the experimental site and the texture on its surface
(Source: Yufan Zhu)

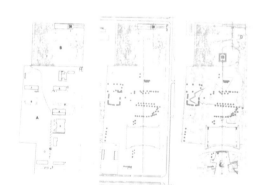

图20 路径的设计一方面契合现有青杨界定的空间秩序，另一方面连接隐形中轴线的两端
（资料来源：朱育帆工作室）

Fig.20 Path designed to fit the new spatial order defined by the poplar trees and connecting two ends of the hidden axis
(Source: Yufan Zhu Studio)

的两个端点形成了一条隐形的中轴线，其中和平之丘既是公园空间的焦点，也是意向的核心。从象征意义而言，和平之丘象征了世界的和平，这是中国发展核工业的终极目标。从空间而言，和平之丘的山顶平台是全园的制高点，沿路径行进的整个过程中人们都可以观看到或者感知到它的存在。设计中有意识地将游人的运动过程与视觉引导方向进行了分离。虽然人们可以直接望见和平之丘，但是却没有办法找到一条直达的通道。相反，人们必须沿着道路蜿蜒前行，一边与先驱者的雕塑并肩而行，一边体会脚下踩踏粗糙的碎石铺地而感受到的疼痛与疲劳，穿过青杨浓密的树叶筛下的跳跃光斑，一路来到下沉广场。在这里年轻物理学家夫妇的事迹正在为空间与造型的语言所讲述（图23，图24）。

叙事性体验过程的基础是路径。与肯尼迪纪念园的路径所不同的是，原子弹纪念园的路径自身就是有形的表达。路径转折和延展的方式暗合中国传统书法艺术的规律、逻辑和美感。在诸如绘画和书法这样的中国传统艺术中，线条以及线形特征是最为精深也最为明显的表现要素。设计师继承了这种文化传统，利用线条形成了风景的表达。这样的线条将新的空间序列与现状杨树界定的空间秩序进行了编织，也把新建的博物馆与公园空间进行了衔接，它赋予了整个方案物质化的形态（图25）。

虽然在这个项目中采用了一种可以与其他文化背景类比的设计策略，而且项目给人留下的印象反映在叙事性特征和空间的创新性表达上，看起来与文化特质关联不大，但事实上，这个项目空间形态的推演与呈现却与根植于中国传统艺术的非物质性特征紧密相连。

of the development of Chinese nuclear industry. Physically, the flat top of the hill was shaped as the highest spot of the entire park, which could be seen or be perceived from all along the path. The visitors' movement was designed to separate from the vision orientation. Although the Hill of Peace was visible directly, it was impossible for the visitors to find a direct path towards the hill. Instead, they had to wind along the path, walking shoulder by shoulder with the pioneers' sculptures, stepping on the rough gravel surface of the path to feel the pain and exhaustion, wandering through the leaping light between the leaves of the poplar trees until the sunken plaza where the young couple's story was revealed with the language of sculpture and space (Fig. 23, Fig. 24).

The whole process of narrative expression was based on the passage. Different from the passage in Kennedy memorial, the passage in unclear memorial park itself is an expression. The way it curves and extends follows the principle, logic and aesthetic of traditional Chinese calligraphy. In traditional Chinese art like painting and calligraphy, lines and linear features were the most important and sophisticated elements to express. The designer inherited such cultural tradition and used a "line" to form a landscape expression. The line woven the new spatial sequences with the space order defined by the existing trees and connected the newly built museum with the space of the park. It gave birth to the physical form of the whole project (Fig. 25).

Although a comparable design strategy of passage to the projects from different cultural background was applied and the project impressed people with its narrative characteristic and renovating spatial expression that seems had little concerns on cultural identity, the project is actually closely connected with intangible features that rooted in the traditional Chinese culture.

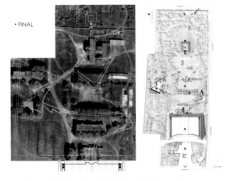

图21 路径的南端连接博物馆的出入口
（资料来源：左图：朱育帆，改绘自谷歌地球地图；右图：朱育帆工作室）

Fig.21 South end of the path connecting the entrance of the museum
(Source: Left: Yufan Zhu, based on Google Earth map; Right: Yufan Zhu Studio)

图22 路径的北端形成纪念公园的核心
（资料来源：Yue Jia）

Fig.22 North end of the path forming the focus of the park (Source: Yue Jia)

图23 路旁的先驱者雕塑（资料来源：陈尧）

Fig.23 Sculptures of the pioneers standing by the path
(Source: Yao Chen)

8.3 汇入洪流：一种立场

"阈""互惠""路径"这些主题都是西方设计理论中的主张。然而，来自中国的设计项目可以在这些主题下与西方的项目相比较，说明所涉及的项目都具有可比性，而且具有相同的背景。这种共同的理论和实践背景正是伴随不可避免的全球化趋势而来。

通过三组比较可见，我们自己的实践可以从物质性和非物质性方面向设计共同的背景贡献独一无二的内涵。这些内涵的外在形式未必与传统直接关联，但是内在的价值却深植于我们特质的文化土壤。保护我们的传统文化非常关键。同样重要的是，需要创造我们自己对于传统文化的当代阐释以形成新的特质，并且在国际化的趋势中去传播这种特质。这样，当我们汇入全球化的洪流时，就有可能对其进行新的塑造。

8.3 In the Cultureshed: a Proposition

Threshold, reciprocity and passage are themes advocated in western design theories. The fact that Chinese projects are possible to be raised to compare with western projects under such themes proves that all the projects involved in the comparison are comparable and share common theoretical context. This context came into being with the inevitable trend of globalization.

Through the three comparisons we found that it is possible for our practices to contribute unique features both tangibly and intangibly to the common context. On such features, the external expression maybe not directly connected to the tradition, but the internal value is deeply rooted in the soil of our cultural identity. It is essential to preserve our traditional culture. And it is equally important to shape new identity by creating our own contemporary interpretation of the traditional culture and to share it within globalization. Then, the flood of the globalization may be reshaped when it immerse more area of our own "cultureshed".

图24 科学家夫妇的故事形成下沉广场的空间主题
（资料来源：贾玥）

Fig.24 Sunken plaza taking the couple's story as the theme of the space
(Source: Yue Jia)

图25 路径的形态暗合中国书法艺术的美，塑造了空间的形态
（资料来源：贾玥）

Fig.25 Path intangibly following the aesthetic of traditional Chinese calligraphy, and giving birth to the physical space of the project
(Source: Yue Jia)

参考文献 / References

BERRIZBEITIA A, POLLAK L, 1999. Inside outside : between architecture and landscape[M]. Gloucester, MA. : Rockport.

FRAMPTON K, 1995. Studies in tectonic culture : the poetics of construction in nineteenth and twentieth century architecture[M]. Cambridge, MA. : MIT Press.

RESTANY P, 1992. Dani Karavan[M]. Munich : Prestel.

SPENS M, 1994. The complete landscape designs and gardens of Geoffrey Jellicoe [M]. New York : Thames and Hudson.

SCHEURMANN I, SCHEURMANN K, 1993. For Walter Benjamin : documentation, essays and a sketch[G]. Bonn : AsKI.

WELLACHER U, 1996. Between landscape architecture and land art[M]. Basel, Boston : Birkhäuser.

陈劲松, 2012. 欧洲音乐沙龙文化的兴起和发展[M]. 昆明：云南大学出版社.

斐莲娜·封·德·海登-林许, 2003. 沙龙: 失落的文化摇篮[M]. 张志成, 译. 台北: 左岸文化.

李雯, 2012. 卡罗·斯卡帕[M]. 北京：中国建筑工业出版社.

一个建筑师关于文化与环境的可持续性的论述

AN ARCHITECT'S DISCOURSE ON CULTURAL AND ENVIRONMENTAL SUSTAINABILITY

国广乔治
建筑学教授
日本国士馆大学

George Kunihiro
Professor of Architecture
Kokushikan University

中文翻译：刘健，李彩歌
Chinese translation by Jian Liu and Caige Li

9.1 前言

1980年代中期以来，中国经历了经济迅速发展，建筑文化变革因而受到特别关注。过去10年，中国建筑师不断学习西方的现代和当代建筑，并开始探索自己的方向。由清华建筑系的创始人梁思成教授开创的传统曾一度中断，现在又重回正轨。日本现代建筑140年的发展可以为今天的中国当代建筑提供参考。

日本作为一个闭关锁国260余年的国家，与中国有着相似的现代化开端。19世纪晚期，德川幕府同意放弃军权，还政于明治天皇，日本自此开始现代化进程。现代化意味着追赶西方及其工业革命。在第二次工业革命中，日本是欧洲和美国之外效仿西方最成功的国家。与工业和所有其他专业领域一样，日本在建筑领域聘请了外国专家。最初，日本并没有建筑师，只有木匠，"建筑"一词是被发明出来的。从那时起，日本的建筑和建筑师逐步建立国际声望，使日本成为当今世界建筑领域的领先国家。

总体来说，包括我在内的建筑师们被教诲相信，"我们是物质环境的创造者，我们知道物质世界是如何建构的，并且了解我们的城市和环境"。虽然建筑师声称自己富有远见，但事实上在公众眼里，我们并非这个世界的领导者，反而似乎是处在整个体系的最底层。最终决定我们生活环境命运的是政治家、经济学家和拥有财富的人，建筑师不过是把他们的决定落到实处。不论建筑师的预期和理想如何，都与社会价值关系不大。在这间演讲厅外，在任何人群聚集的场所，即便是那些普利兹克奖得主，也少有人了解建筑师们在思考什么，更不必提及他们的名声和功绩。

9.1 Introduction

Transformation of architectural culture should be of particular interest in China, since there is a rapid development of economy beginning in the mid-1980s to the present. During this period, Chinese architects have studied the modern and contemporary architecture of the west and have begun to develop their own direction in the last decade. The tradition begun by Professor Liang Sicheng, the founder of the architectural program at Tsinghua University, was disrupted for some time, but it is back on track again. The 140 years of development of modern architecture in Japan can be a reference to the course of Chinese contemporary architecture today.

Japan's modernization had a similar beginning, coming out of the 260-year isolation from the rest of the world. The nation began her modernization in the late 19th Century, when Emperor Meiji was restored to the throne after the Tokugawa Shogunate agreed to forfeit its military rule. Modernization meant catching up with the west and its Industrial Revolution. Japan was the most successful country outside Europe and the United States to emulate the west during the second Industrial Revolution. In architecture, as in all other fields of specialization and industry, Japan contracted "oyatoi-gaikokujin", or the "hired foreigner", who were experts in their field of specialization. To begin with, there were no architects in the country, only master carpenters. The word "architecture" had to be invented. Since then, Japanese architects and architecture had built up an international reputation, making Japan one of the leading countries in the world of architecture today.

Generally speaking, architects, including myself, are taught to believe that "we are the creator of our built environment and we are the ones who know how the physical world is structured and understand about our cities and our environment". Although architects claim to be visionaries, in reality, we are not the one to lead the world in the eyes of the public. On the contrary, architects seem to be at the bottom of the hierarchy. It is the politicians, the economists, people with wealth who decide on the fate of our living environment. At the end of the day, architects provide whatever they decide should be in place. It seems that whatever architects envision and dream are less relevant

9.2 生活经历与双重文化

9.2.1 开端

今天，我的演讲有关一个建筑师的生活与其建筑文化发展之间的关系，希望能为同学们规划个人事业提供参考。这是一个关于我自己事业发展的故事。相信你们听过许多中国和国外建筑师的演讲，我想大部分建筑师都会展示他们的项目和成就，但少有谈论他们个人的故事。我想换种方式来介绍自己，结合我的人生历程来介绍我的工作，以揭示一个人的生活及其创意工作之间的关系。

我的人生历程涉及了两个大洲的诸多城市。我出生于东京，孩提时代生活在美国密歇根州底特律市，之后回到东京完成小学。几年之后，全家搬到美国西海岸的旧金山；那里有很多中国人，我在那里开始了解中国文化。在后来的人生历程中，我先后在波士顿、洛杉矶和纽约生活；目前，我居住在东京，并在亚洲不同地方工作。

故事要从我的祖背讲起。我的爷爷奶奶是从日本移民到美国的农民，他们20世纪初在加州相见，经过一番打拼，开始在加州奥克兰经营一处苗圃。我父亲是他们四个子女中的长子，他5岁时全家决定迁回日本。这是我爷爷奶奶做出的一个不同寻常的决定，因为那时的移民通常不会回到他们的母国。然而我爷爷看准机会，要在美日之间开展国际业务，所以决定离开加州。于是我父亲作为一个日裔美国人，在东京度过了他的青春时代。不幸的是，"二战"爆发，国广家族因为生活在"敌对"国家，全部失去美国公民身份。

战争结束后，我父亲为了重新获得美国公民身份，前往美国驻日军队工作，并在10年后如愿以偿。在当时

to the values of the society. Outside this lecture hall, or in any gathering, including those of the Pritzker Prize winners, a very few has knowledge of what architects are thinking, let alone their fame and achievement.

9.2 Life Journey and Bi-Culture

9.2.1 The Beginning

Today, I would like to speak to you about the relationship between the activities of an architect and the development of one's architecture culture. I hope that the content of my presentation can contribute as a reference to the way a student plan his/her personal career. This is a very personal story of my career. I am sure that you have heard lectures by many architects, both Chinese and foreign. I would say that most architects present their projects and achievements, but they seldom relate about their personal saga. I would like to represent myself differently. I will present my work in relation to my life journey, hopefully, to illustrate the relationship between one's life and the creative work.

My life journey centers around several cities in two Continents. I was born in Tokyo, had lived in Detroit, Michigan, as a child, and returned to Tokyo to complete my elementary education. Several years later, our family relocated to the West Coast of the United States, to San Francisco where there is a very large Chinese population. It was in San Francisco that I was introduced to the Chinese culture. Elsewhere during the course of my life, I had lived in Boston, Los Angeles and New York. Today, I am based in Tokyo and working around the Asia Region.

The story begins with my grandparents. My grandparents were immigrant farmers from Japan, who met in California at the beginning of the 20th Century. They worked their way into owning a nursery business in Oakland, California. My father was the eldest of four children. When my father was five years old, the Kunihiro family decided to move back to Japan. This was an unusual move by my grandparents. Immigrants in those days didn't usually return back to their native land. However, my grandfather saw an opportunity to start an international business between the U.S. and Japan and he decided to leave California. Hence, my father grew up as a Japanese-American in Tokyo.

的日本，为美军工作能够拥有一些特权；我父亲可以购买美国生产的汽车，可以在美军基地的邮政交易商店购物，买到各种美国产品。我有幸在这样的环境下长大。1950年代，孩提时代的我经常可以到东京市中心区的美式免下车餐厅享受汉堡、薯条和冰淇淋，我父母还带我去东京的主题公园，享受登上颇具未来感模拟飞机的乐趣。从长远看，孩提时代的类似经历给我以及与我同龄的孩子们带来了对未来的希望和梦想。

那段时间，我父母正在计划移民美国；为了获得旅行证件，他们不得不解决我的公民身份问题。当时，我父亲已重获美国公民身份，我母亲也得到了证明文件，可以在抵达美国后成为美国公民；而我出生在日本，有一个美国籍的父亲和一个日本籍的母亲，这意味着我在日本家庭登记中无法获得正式身份。最终，我父母通过红十字会获得了我的旅行证件——身份是"孤儿"！在那之后的人生中，我不断面对身份危机：我是日本人还是美国人？当时，为了克服这个危机，我将目光投向一个更大的版图，决定把自己认作"亚洲人"。

9.2.2 移居到美国

1958年，我们全家搭乘螺旋桨飞机离开东京前往密歇根州，从一个亚洲城市落户底特律郊区。那个社区的大多数居民是波兰裔的美国人，亚洲人寥寥无几，只有一家韩国人经营的亚洲商店，而我也是学校里唯一的亚洲学生。当时我并不知道，其实底特律市中心有一个小型中国社区。在底特律，我再次经历启蒙。底特律有一座桥跨越底特律河，通往对岸的加拿大温莎市。对一个来自岛屿国家的孩子来说，坐在车里通过一座桥梁，跨越国界进入另

Unfortunately, the World War II broke out and the Kunihiro family, all American Nationals, lost their citizenship as they had lived in the "enemy" territory.

After the end of World War II, my father went to work for the U.S. Military Occupation Force, in order to regain his lost U.S. citizenship. After ten years of working for the U.S. Forces, my father successfully regained his American citizenship. Working in the American Military Services had certain privileges in Japan in those days. My father could afford an American made car, shop at the Post Exchange on the military base, where American products were readily available. I was fortunate to grow up in such environment. As a child in the 1950s, we frequently had burgers, fries and ice cream at the American-style drive-in restaurant in the central district of Tokyo. My parents also took me to the local theme park where I enjoyed futuristic airplane ride. In hindsight, such childhood experience gave myself and the children of my generation hope and dream for the future.

Around this time, my parents were planning to move to the U.S. Again, they had to sort out my citizenship situation for my passport to travel. My father regained his citizenship; my mother received her documents to travel to the U.S. to get her naturalized citizenship. On the other hand, I was born to an American father and a Japanese mother in Japan. What this meant was that I did not have an official status in the Japanese family registry. Finally, my parents got me a travel document issued by the Red Cross as an "orphan"! I was a boy without a home country. As a result, I faced an identity crisis later in my life. Am I a Japanese or an American? At the time, I looked at a larger picture to overcome this personal crisis. I decided to identify myself as an "Asian".

9.2.2 Moving to the United States of America

In 1958, our family left Tokyo on a propeller plane and headed to Detroit, Michigan. From an Asian city, we ended up in a Detroit suburb where the residents were mostly Americans of Polish decent. I remember that there was only one Asian market which was owned by a Korean family and very few Asians around my community. In fact, I was the only Asian student in my school. At the time, I was not aware that there was a local

外一个国家，这是一种全新体验，一种有关空间、时间、文化的启示，对当时的我产生了重要影响。

在底特律生活了大约一年后，我重新回到东京。我很感谢我父母在我小的时候把我带到美国学习英文；事实证明，这个技能非常有用，帮助塑造了我的人生。回到东京，我在日语环境里完成小学教育。

在我从私立小学毕业后，又到了搬家的时候。1964年，在东京举办奥运会那年，我离开日本前往加州，在那里度过了我的青少年时代。1964年的东京就像1988年的首尔和2008年的北京，洋溢着奥运会带来的热烈氛围。这对日本而言意义重大，因为它寓意着日本在战败后的复兴；在广岛和长崎遭遇原子弹爆炸19年之后，日本以一个重获新生的姿态回到国际社会。我相信韩国在1988年和中国在2008年举办奥运会时，也有同样积极乐观的情绪；但在日本，成功举办奥运会的欢乐其实加剧了人们对过去的痛苦记忆。日本原本准备举办1940年奥运会，却因日本军队入侵中国、直至卷入第二次世界大战而作罢；因此，1964年的东京奥运会标志着日本重回国际舞台。遗憾的是，我在东京奥运会召开前5个月的5月31日离开日本，最终错过了所有的庆典活动。

9.2.3 成长在旧金山

旧金山是一座宜居城市，既有文化城市的氛围，又有适宜的尺度规模，人口接近100万。这座城市建在群山河谷起伏多变的土地上，被加州良好的自然景观所环绕。多种族混居使旧金山成为一座熔炉，主要人口是西班牙语系的居民，还有大量亚洲人，包括中国人、日本人、韩国人、菲律宾人和越南人。

Chinese community in downtown Detroit. In Detroit, I had another enlightening experience. There is a bridge connecting to our city with Windsor, Canada across Detroit River. For a child from a country made up of islands, continuous travel on a car over a bridge to cross a national border was a new experience. It was a space-time-culture revelation which had for certain an impact on me at the time.

My next re-location was back to Tokyo. I was in Detroit for approximately one year. I am thankful to my parents for putting me in the American environment to learn English early in my life, because this skill became useful and helped shape my future. Back in Tokyo, I spent my remaining elementary school days in the Japanese environment.

Upon graduating from a private elementary school, it was time to move again. In 1964, the year of the Tokyo Olympic Games, I left Japan for California where I will spend the rest of my teenage days. The 1964 Tokyo, like 1988 Seoul and 2008 Beijing, had an air of excitement with the Olympic Games. It was significant for Japan, as it had overcome the defeat in the War. Nineteen years after the atomic bomb had dropped in Hiroshima and Nagasaki, Japan returned to the international community as a fully rejuvenated nation. I am sure that Korea in 1988 and China in 2008 experienced similar optimism hosting the Olympics. In the case of Japan, the bitter memory of the past had even enhanced the joy of hosting a successful Olympic Games. The nation was preparing to host the 1940 Olympic Games, when, Japanese Imperial Military invaded China which evolved into an involvement in World War II. Thus, Japan was resurrected in the Post World War International Community. Unfortunately for myself, I missed all the festivities of the Tokyo Olympic Games, as I left Japan on May 31st, five months before the opening of the Games.

9.2.3 Growing Up in San Francisco

San Francisco is a great city to live in. The city has an aura of a cultural city with just a comfortable size, with a population of less than a million. The city is built on a varying terrain with hills and valleys, surrounded by fine California natural landscape. Ethnic mix makes San Francisco a melting pot. There is significant population of Spanish speaking community, a large

定居旧金山后，我进入当地一所公立中学读书。和任何孩子初到陌生环境时一样，我面对结交新朋友、融入新社区的挑战。当地有一群日裔美国孩子，可惜他们认为我是"刚下船的"FOB，不肯接纳我进入那个群体，白人和非洲裔美国人对于一个矮小的日本孩子来说过于高大；所以，我最终去找了邻居的中国小孩和菲律宾小孩，受到了他们的欢迎。这是我生活中的一个重要转折，在这个旧金山的社区里，我开始了与中国社区维系一生的关联。

1970年高中毕业，我进入加州大学伯克利分校。这一年，正在越南酣战的美军空袭了柬埔寨，引发了遍布全美的学生抗议。我的母校作为美国学生运动的带头学校之一，与哥伦比亚大学共同领导了全国性的学生抗议活动。校园里的史鲍尔广场上时有集会发生，我在环境设计学院的第一个学期就多次参加，支持学生抗议活动的教授们也经常在当地的教堂或者公园上课。

9.2.4 进入建筑学院

高中时，我一直希望进入太空工程领域，为美国的太空项目工作，且对此信心十足。得益于日本高水平的小学教育，我在数学方面表现优秀，虽然数学成绩在日本只能算平均水平，但在美国的初中和高中一直名列前茅。然而高中最后一年，在被加州大学伯克利分校录取之后，我得了一个D的成绩。我忽然意识到自己并不适合当工程师，以往的梦想就此戛然而止！一位亲戚看到我在艺术和科学方面的兴趣，向我提议学习建筑；于是我迅速改变志愿，转向建筑。

Asian population, consisting of Chinese, Japanese, Koreans, Philippines and later on, the Vietnamese.

Our family settled in San Francisco and I attended a local public middle school. As any new kid in town, I had a challenge of making new friends to become a part of the community. I began looking for new friends. There was a group of Japanese-American children in the neighborhood. Unfortunately, these children saw me as a boy "fresh off the boat", or the "FOB's". I was not able to fit into this group. Caucasians and African Americans were a sort of oversized for a short Japanese boy from Tokyo. Finally, I looked to the group of neighborhood Chinese and Pilipino boys and was welcomed into the group. This was a turning point in my life. A life-long relationship with the Chinese community began here in a San Francisco neighborhood.

In 1970, I graduated from a public high school to enter the University of California at Berkeley. This was the year when the U.S. Military, fighting in Vietnam, launched air strikes in Cambodia. Student protest was mobilized nationally, and my university which was one of the leading universities in the student movement, led the national student protest along with Columbia University. Gatherings at the Sproul Plaza on campus was a regular scene. My first semester in the School of Environmental Design was going to many of these gatherings, attending classes in places like a local church, or in the park, where the professors sympathetic to the protest held the classes.

9.2.4 Entering Architecture School

Up until my senior year in high school, I had decided to pursue a career in the space industry, hoping to be working in the U.S. space program. I was confident, because my grades in mathematics were high, as a result of the level of elementary education in Japan. Although I had an average grades in Mathematics course in Japan, I became a regular honor roll student in middle and high schools in the U.S. However, such dream came to an end, when I received a dismal "D" in my senior year, after already been admitted to the University of California at Berkeley. This was a moment when I realized that I was not cut to become an engineer, making a quick shift to architecture. The decision came as I took up on a suggestion

9.2.5 美国征兵

在此，我还想讲一下越战期间参加美国征兵的故事。当时的中国正在经历"文化大革命"，而美国则深陷印度支那战争，年满19岁的年轻人都被要求参加征兵。征兵过程像乐透彩票，由系统抽取生日决定，通过电视在全国播出，其中前60个被选中的年轻人会被派往越南。我当然不想被送往越南，与我的亚洲同胞作战，因为我很清楚自己的亚洲血统。为了逃避兵役，我曾考虑离开美国去往加拿大，或者有意破坏自己的健康以被视为不适合兵役。最终，我还是决定去碰碰运气。当征兵电视转播结束时，我所有的担心终于化解；我的生日在非常靠后的315号被抽中！这是我生命中的一次危机，以至我对这个神奇的数字记忆犹新。

回首往事，越南战争是一场恐怖的灾难，而不是一场正义的战争。我在建筑生涯中结识不少越南建筑师，他们在年轻时都曾经历过那场可怕的战争；我曾多次与他们沟通，相互了解各自参与战争的经历。其中一位同事在当年的新年攻势时就居住在岘港，在岘港山上亲眼看到美国海军舰队向城市开炮；我和他并肩站在那座山顶时，他给我讲述了那个可怕的场景。另外一个故事来自我一个学生的父亲，他是北越军队的将领，把自己的经历讲给儿子。南亚深陷战争时，我在伯克利校园里参加学校的反战抗议活动，向当局投掷石块，以示一位生活在美国的亚洲人的抵抗。

9.2.6 参加亚洲社区的社区设计

在大学，亚洲建筑协会（AAA）让我有机会和其他亚洲学生见面和共事。我和协会的同事在校外的亚洲社区

made by our family acquaintance who saw my interest, both in art and in science.

9.2.5 The Military Draft

I would like to speak a little more about the military draft in the United States during the Vietnam War. In China, it was a decade of the Cultural Revolution and, in the U.S., it was the war in Indochina. Each year, all the eligible nineteen-year old youths were subject of the draft. The youths were selected according to the result of a lottery, picked by one's birthday, broadcasted on national television. I remember that if one's birthday was selected in the first 60 draws, those youths were Vietnam-bound. I did not want to be sent to Vietnam to fight my own Asian people. I had a very clear awareness about my Asian heritage. Just prior to the day of the draft lottery, I considered resisting the draft by leaving the U.S. for Canada, or intentionally damaging my health to be labeled unfit for the military. In the end, I chose to try my luck in the lottery. When the television draft lottery program ended, all my worries had passed. My birthday was drawn towards the very end at number 315. Since it was one of my life crisis in my life, I still remember this magic number.

Looking back, the Vietnam War was a frightening disaster and it was not a war to be justified. It is ironic that, through my career in architecture, I have come to know architects in Vietnam, who had experienced the war in their youth with horror. I had several occasions with these foreign colleagues relating to each other about our own experiences on either side of the war. One colleague, who lived in Danang area during the Tet offensive, had seen from atop the hills the U.S. Naval ships shooting canons towards the city. I stood with him on the very hill as he was relating this story. Another counterpart was the father of one of my students. His father was a North Vietnamese army general and related his experiences to his son. While the war was going on in Southeast Asia, I was involved in the anti-war protest on campus of my university, throwing rocks at the authority as a symbol of resistance as an Asian living in the U.S.

9.2.6 Participating to Community Design in the Asian Community

In school, the Asian Architecture Association (AAA) gave

共同建立起一个公益性社区设计小组，成员包括来自中国、韩国、菲律宾、日本和印度尼西亚的学生，致力于运用在建筑学院学到的知识和技能，为当地亚洲社区提供帮助。亚洲邻里设计（AND）由此诞生。我们有幸得到某基金会提供的启动资金，帮助日本城的一个店主设计店铺门面，用来与当地的开发机构讨价还价。我们还为日本城里一片被指定拆除的房屋编制了更新计划，使得年长的亚洲居民能够保留自己的住所；按照原有规划，这片房屋被拆后将在原址建设公寓，必然带来地价攀升，迫使租住于此的第一和第二代移民搬迁，离开他们熟悉的生活环境。另外，这些亚裔美国人居住的房屋多属维多利亚风格，作为当地文化遗产具有重要价值，但在1970年代，并未被当地社区所理解和重视。

多年以后的2009年，当我再次回到旧金山湾区时，发现AND依然还在，并且成为湾区非常重要的公益性社区组织。克林顿总统当政时，曾邀请AND介绍其就业培训项目，并对其付出的努力和取得的成就大加赞赏。一个由学生发起的社区设计组织，在40多年后仍在当地社区里为亚裔美国人提供帮助，作为其发起成员，我对此感到骄傲和高兴。

9.2.7 选择建筑设计作为信仰

大学期间，除了对亚洲社区充满热情外，我还探索了建筑设计的梦幻世界。旧金山既有富有地方传统的湾区风格建筑，又有受美国东海岸影响的现代主义建筑。当时最时髦的项目是内河码头都市更新，尤其是约翰·波特曼设计的凯悦酒店及其创造的中庭空间原型（图1）。波特曼既是一位建筑师，还是一位开发商；建筑师加开发商的双

me an opportunity to meet and work with other Asian students. I joined my AAA colleagues in the effort to establish a non-profit community design group outside of the university, in the local Asian communities. Our group consisted of Chinese, Korean, Pilipino, Japanese and Indonesian students. We were all committed to offering our learned skills in architecture to the local Asian communities. Thus, Asian Neighborhood Design (AND) was born. Fortunately, our group was able to secure funding from a foundation to begin our operation. AND assisted a shop owner in the Japan Town by preparing a design alternative for their storefront to be used to negotiate with local redevelopment agency. We also designed a renovation plan for houses slated to be demolished to make a way for the construction of housing development, which would raise the land prices in the neighborhood, making the elderly Asians to keep their homes. First and second generation immigrants renting homes in the Japan Town would be evicted and relocated from their familiar environment, as a result of the redevelopment. Moreover, the Victorian style houses, that these Asian Americans lived, were valuable as local cultural heritage, not fully understood nor appreciated by the community in the 1970s.

Many years later, in 2009, I returned to San Francisco Bay Area to learn that the AND is still in existence and has grown to become a very important non-profit community organization in the Bay Area. President Clinton, when he was in office, invited the AND to make a presentation on the AND's job training program and commended the organization's efforts and accomplishments. I am proud and happy to know that a student initiated community design group, in which I was a founding member, is still helping the Asian Americans in the local community after more than forty years in service.

9.2.7 Choosing Architecture Design as Religion

Besides my passion for the Asian community, I also discovered the fascinating world of architectural design. We were fortunate, in San Francisco, to have both the local tradition of the Bay Area style architecture and the influence of modernism coming from the East Coast. The latest, at the time, was the Embarcadero Redevelopment, featuring Hyatt Regency Hotel (Fig. 1), designed by John Portman. Portman was an architect

图1 波特曼在凯悦酒店设计中创造的中庭空间原型
（资料来源：George Kunihiro）

Fig.1 The prototype of grand atrium created by John Portman at the Hyatt Regency Hotel
(Source: George Kunihiro)

重角色，对于一个大学三年级的建筑学生来说充满新奇。

位于旧金山以北180千米的海洋牧场公寓项目（图2），由查尔斯·摩尔、威廉姆·特恩布尔、约瑟夫·埃什立克、劳伦斯·哈普林等多位知名建筑师和景观建筑师合作完成。美国建筑师查尔斯·摩尔曾在加州大学伯克利分校执教，后担任耶鲁大学建筑学院院长，他的建筑颇受美国自然景观和社会价值的启发。海洋牧场项目中的公寓建筑采用了美国西部的本土建筑材料，成为他在那个时期的代表之作。项目场地面向太平洋，气候条件恶劣；树立在公寓旁边的雪松不仅很好地适应了当地的气候条件，并与当地的景观特质相得益彰。著名景观建筑师劳伦斯·哈普林在充分的场地调查基础上，在1960年代中期完成了项目的总平面设计。

9.2.8 前往波士顿

从一个优秀的研究导向的建筑学院毕业后，我想去一个设计导向的建筑学院继续深造。于是，我申请了哈佛大学设计学院（GSD），并有幸被录取。那是我第一次前往美国东海岸生活。就像中国的北方和南方之间存在地区竞争一样，美国的东海岸和西海岸之间以及南部和北部之间也存在着地区竞争。抵达GSD，我发现自己成了学院里的极少数，我的同学对于来自西海岸旧金山的亚裔美国人充满好奇。当时，学院里还有些外国学生，所以又一次，我作为亚洲人加入中国人和华裔美国人的团队。

GSD系馆由约翰·安德鲁设计，1973年建设完工。这个有趣的空间像一个庞大的厂房，容纳了阶梯状的楼层，造就了一种社区感。在这里，学生们可以自由走动，与设计工作坊里来自其他系的学生交流。在GSD求学期

who created the prototype grand atrium lobby space in his Hyatt hotels he was designing at the time. He was also a developer, a role, which was something new for a third year student, an architect-developer.

Another example is Sea Ranch, a development 180 kilometers north of San Francisco, which was designed by a group of prominent architects and landscape architects, including Charles Moore, William Turnbull, Joseph Esherick and Lawrence Halprin. Charles Moore was an American architect who taught at the University of California at Berkeley, and later became a Dean at Yale University. His architecture was inspired by the American vernacular and the values of the American society. The condominiums at Sea Ranch, using the indigenous materials of the American West, was his representative project during that period of his career (Fig. 2). Local climate in Sea Ranch is rather harsh with the land fronting the Pacific Ocean. The cedar siding weathers well to the local climate and blends into the landscape. Lawrence Halprin, the well-known landscape architect, surveyed the property and completed a master plan in the mid-1960s.

9.2.8 Going to Boston

After graduating from an excellent architecture program with leaning towards research, I looked for a design-oriented program in the graduate school. I had decided to apply to the Harvard Graduate School of Design (GSD) and was fortunate to be accepted into the program. It was my first time living in the East Coast of the U.S. Like the regional rivalry between Northern and Southern China, U.S. also has a rivalry between the East Coast and the West Coast, and between the North and the South. When I arrived in the GSD, I found that I was a minority. My classmates were curious about a West Coast Asian American from San Francisco, a very minority in the School. We did have a number of foreign students and, here again, I had become an Asian to join the group of Chinese and the Chinese Americans in the School.

The Gund Hall, designed by John Andrews and completed in 1973, was an interesting space to study, with the stepped levels of the floors under a large single factory-like space. It was a space which induced a feeling of community. The students freely moved about the space to meet students from other departments

图2　海洋牧场项目中的公寓建筑所表达出的质朴的乡土建筑风格
（资料来源：George Kunihiro）

Fig.2　The Vernacular and non-pretentious architecture expressed by the condominiums at Sea Ranch
(Source: George Kunihiro)

间，我的指导教师包括了曾在柯布西耶工作室就职的约瑟夫·扎勒斯基、设计了波士顿市政厅的杰拉德·科曼，以及其他知名建筑师和学者。另外，GSD还通过讲座和期末评图，让学生们有机会结识世界知名建筑师，其中就包括日本建筑师桢文彦。桢文彦因为1960年在东京召开的世界设计大会上的发言而受到关注，他也是GSD毕业的学生，我向来尊敬他的设计和写作。几年前他在伯克利任职访问学者时，正在读本科的我曾与他谋面，但并未有机会交谈。在他到哈佛大学讲座之际，我径直找到他，希望他给我一个面试的机会，他慷慨地答应了我的请求。第二年夏天，我飞往日本拜见桢文彦并参加面试。可惜1970年代中期，日本正在经历石油危机，他的事务所无法雇佣一位年轻的毕业生。从哈佛毕业后，我回到旧金山湾区，决定首先解决自己的财务问题，在一家出租车公司获得一个出租车司机的职位。靠开出租来维持生活的经历虽然短暂，但乐趣无穷；通过开车时与乘客聊天，在旧金山的大街小巷穿梭不停，使我洞见了城市社会的复杂。我有时甚至感觉我拥有了整座城市！

9.2.9 辗转洛杉矶和纽约

在我的求职申请被桢文彦事务所委婉拒绝之后，我先后在旧金山本地的几家设计导向的建筑公司工作。1979年，我南下搬到了美国当时的第三大城市洛杉矶。洛杉矶地势平坦、地域广阔，主要出行方式是汽车；1970年代，这里除了公共汽车之外，没有任何其他形式的公共交通。在这里，除了迪士尼、好莱坞、一些油田以及南加州特有的海滩风光外，没有东海岸常见的"文化"氛围。南加州的建筑传统直到20世纪伊始才开始出现，代表人物

in the studios. During my years at the GSD, my professors included Joseph Zalewski who had worked in the atelier of Le Corbusier, Gerhard Kallman who designed the Boston City Hall and other prominent architects and academics. In addition, the GSD made sure that we were exposed to the world reknown architects in the form of lectures and final jury at the end of the school year. One such architect was Mr. Fumihiko Maki, up-and-coming Japanese architect already making his marks in the World Design Conference held in Tokyo in 1960. He was a graduate of the GSD and I always respected his work and his writings. I had seen him as an undergraduate student in Berkeley, when he was a visiting scholar a few years earlier, although I did not have an opportunity to speak to him. At the lecture in Harvard University, I went up to him to ask for an interview. Mr. Maki was generous to give me a positive answer. In the following summer, I flew to Tokyo to meet Mr. Maki for an interview. After the interview, Mr. Maki advised me to wait for further communication from his office regarding the position. Unfortunately, Japan was experiencing the so called "Oil Shock" in the mid-1970s and the office was not in the position to hire a young graduated at the time. In the meantime, I had graduated from GSD and returned back to the San Francisco Bay Area. Back home, I decided to keep concentrate on my financial situation. I got a job at the taxi company as a cab driver. Driving for living was short but a very interesting experience. In hindsight, the experience gave me an insight into the dynamics of urban society, making conversations with my customers while driving in and around the streets of San Francisco. I felt, at times, like I owned the City!

9.2.9 Heading to Los Angeles and New York

After being politely turned down by the office of Fumihiko Maki for a junior architect's position, I went to work for several local design-oriented architectural offices in San Francisco. In 1979, I made a move down the coast to Los Angeles, the third largest city in the U.S. at that time. Los Angeles is flat and vast. Main mode of transportation is automobile. Public transportation, except for buses, were non-existent in the 1970s. I felt that there was not much of "culture" as we knew in the East Coast then, except for Disneyland, Hollywood, a bit of oil fields and the California beach scene indigenous to Southern California.

包括Green & Green的建筑师及其开创的新艺术/日式风格建筑；之后是西班牙风格建筑，以及由《艺术与建筑》杂志赞助的实验性住宅项目。1970年代，弗兰克·盖里的建筑使洛杉矶名声鹊起，由汤姆·梅恩和迈克尔·罗东迪合伙经营的国际知名设计公司莫菲西斯开始在当地生产一些小型的重要建筑项目，建设中的高速路立交和工厂机械的景象都带给汤姆·梅恩以灵感，他最终也因其对机器图像的追求而获得普利兹克奖。

在洛杉矶，我首先去一家荷兰裔本地建筑师开设的事务所学徒，他的设计灵感源于1980年普利兹克奖得主、富有传奇色彩的墨西哥建筑师路易斯·巴拉甘。两年后，我在加州市中心开设了自己的事务所。5年后，在成功帮助一家日本企业在曼哈顿开展业务之后，我在纽约设立了事务所。纽约是一座具有巨大影响力的国际都市，生活在纽约一直是我的梦想；我在哈佛求学时曾多次前往纽约，总是陶醉于那里浓厚的都市生活气息。许多具有国际影响力的建筑师，如菲利普·约翰逊、伯纳德·屈米、贝聿铭等，都在曼哈顿设有办公室。我的小办公室就与另一位国际知名建筑师彼得·埃森曼的事务所隔街相望。

我在纽约主要承接日本项目，因为日本的泡沫经济带给建筑设计领域无限活力。但最终泡沫破灭，来自日本的项目委托戛然而止。当时，经济学家和投资者预测，中国将在下个世纪迅速崛起，亚洲将会成为热门地区；美国随处都能听到"21世纪将是亚洲的世纪"的预言，英国正在准备把香港归还中国，空气中弥漫着激动的情绪。于是，我做出了人生中又一个重要决定，关闭了我在纽约的事务所，回到日本！

Architectural tradition in Southern California began at the turn of the Century with architects like Green & Green and their Art Nouveau/ Japanesque style. Spanish style architecture, the Case Study Houses, the experimental residential architecture sponsored by the Arts & Architecture Magazine followed. The 1970s saw Frank Gehry putting Los Angeles on the map. Morphosis, a partnership of Thom Mayne and Michael Rotondi, now an internationally recognized firm, was just beginning to produce small but important buildings locally. Images of the freeway interchanges under construction and the machinery of the factory became inspiration to Thom Mayne in his "deadtecture". He eventually received a Pritzker Prize for his pursuit of machine imagery.

In Los Angeles, I apprenticed at the Santa Monica office of a local architect of Dutch origin, whose inspiration was Louis Barragan, a legendary Mexican architect who won the Pritzker Prize in 1980. Two years later, I was ready to start my own practice in downtown Los Angeles. Five years later, I opened my office in New York, after assisting a Japanese entrepreneur in starting his business in Manhattan. It was one of my dreams to live in New York, because it was the international city of great influence. I had visited New York City several times during my Harvard days and was always fascinated by the intensity of its urban life. Internationally influential architects like Philip Johnson, Bernard Tschumi, and I. M. Pei, had their offices in Manhattan. My small office was across the street from Peter Eisenman, another influential architect of the generation.

From my New York studio, I worked on projects in Japan, where the bubble economy energized the world of architecture and design. Eventually, the bubble burst and the commission from Japan stopped flowing in. Nevertheless, Asia was going to be the hot place, according to the economists and the investors whose keen sense of global shift predicted the rise of China in the coming Century. "Twenty First Century will be an Asian Century" was the prediction we heard in the U.S. Great Britain was preparing to return Hong Kong back China and the excitement was in the air. I made a major decision of my life, and closed my office in New York and headed to Japan!

9.2.10 离开美国、回到亚洲

从我在东京开设事务所和开始教书，至今已有20年时间。正如经济学家和专家们预测的那样，在此期间，亚洲成为全球经济中心。当美国和欧盟的发展踟蹰不前时，亚洲，特别是中国，成为世界经济的救世主。20年里，我在亚洲的不同地方旅行，进行田野研究，针对不同文化及其面临的挑战积累了足够的一手经验。今天，我虽然回到我的母国日本，但我的视野已经扩大到覆盖了整个亚洲。亚洲建筑师协会（ARCASIA）是由亚洲地区21个国家级建筑机构组成的区域性组织，我作为该协会2011—2012年的主席，有机会成为亚洲建筑师在全球舞台上的代表。两年中，我不断旅行、出访，曾在一周里走过戛纳、巴黎、丹吉尔，代表ARCASIA分别与国际建筑师协会（UIA）、欧洲建筑师协会（ACE）和非洲建筑师协会（AUA）的主席见面。

在我看来，亚洲并非始于日本，而是东亚；韩国和日本都受中华文化的强烈影响；它们虽有不同，但我相信"一个亚洲"的理念。东亚之后还有东南亚，越南同样受到中华文化的影响；本土文化传统叠加殖民文化传统构成当代社会的财富，使我们可以从当地传统中学习提高可持续建筑能源利用的有效途径。之后还有南亚，古老的印度传统以及源自欧洲各地的殖民影响，共同造就了新兴的世界经济枢纽。亚洲拥有丰富的人文和文化资源，它们之间的互动必将在新世纪引领世界发展。

9.3 建筑作品中的双重文化影响

下面我想说说我的作品，其中包含了双重文化的影

9.2.10 Leaving the U.S. to Asia

It has been twenty years since I set up my practice and began teaching in Tokyo. As the economists and experts had predicted, Asia has become the center of global economy. As the U.S. and the European Union faltered, Asia, especially China, has become the savior of world economic system. During this period, I have been travelling and conducting field research in various parts of Asia and I have accumulated enough experience to learn firsthand various cultures and their challenges. Today, I have returned to my home environment of Japan, but my field has expanded to cover the entire Asia region. As the 2011-2012 President of Architects Regional Council Asia, or ARCASIA, which is a council of presidents of 21 national architectural institutes of Asia region, I have had an opportunity to represent the Asian architects in a global arena. In the two years, I have traveled vast distances. For example, I once took a weeklong trip to Cannes, Paris, and Tangiers, where I represented ARCASIA to meet with the Presidents of the International Union of Architects (UIA), Architects Council of Europe (ACE) and the African Union of Architects (AUA) respectively.

My personal "Asia" goes beyond Japan. It begins with East Asia. China, Korea and Japan have a common heritage with a strong influence from the Chinese culture. Although there are differences, I believe in the spirit of "One Asia". Beyond East Asia, there is Vietnam with influence from China. Southeast Asia, with indigenous cultural heritage overlaid with history of colonial rule can be an asset in our contemporary society, because we can learn about the solutions to energy efficient lifestyle in a sustainable architecture from the local tradition of the region. South Asia is another region where the ancient Indian heritage and the colonial influences from various parts of Europe merged to create the emerging world economic hub. Thus, Asia has human and natural resources yet to be consumed. Having so many cultures interacting in Asia, our region, indeed, will lead the world in this Century.

9.3 Bi-Cultural Influence in Architectural Works

Now, I would like to present my own work. I think it will

响，相信你们一定会感兴趣。在我的潜意识里，东西方的理念始终并存。日本文化根植于若干宗教教义，包括源自本土的神道教，来自中国的佛教，以及引自西方的基督教，它们和平共存，甚至共存于一个家庭的生活中。例如，一个家庭可能在厨房供奉神道教的神仙，在卧室用佛龛供奉祖先，把孩子送到天主教幼儿园，于是造就了日本人的双重性格，即理性与精神的矛盾统一。作为一个具有双重文化特点的建筑师，我想通过我的作品对此加以说明。

9.3.1 罗莎别墅

这个项目是位于长野县山区度假胜地——轻井泽的一座夏日别墅，距离东京约1小时新干线车程。建筑被设计成一个边长8.5米的立方体（图3），坐落在松树环绕的狭小场地上。设计目标是使其成为使用者从高强度的城市生活进入宁静安逸的自然环境的转换空间，从迈进场地、走上坡道，到进入房间、走下台阶，最终来到客厅，形成完整的运动序列，然后跨过门槛进入自然空间。在这个白色盒子里，城市生活长期积累的负能量得以洗涤一空，使用者每次到访都可得到"重生"。

建筑造型采用了组装而成的盒子形态，用以表达人工几何体量被"解构"的过程，像人体的消化器官消化食物一样，使建筑体量被解构之后消融在自然环境里；与此同时，使用者的精神也可以在自然环境里得到重生。这个项目同时蕴含了理性和精神要素，白色的建筑在冬天可以融入白雪，在夏季可以反射阳光；这既是理性思考的结果，事实上也给使用者带来精神上的体验。

这栋建筑遵循当代空间理念设计而成，拥有明确的

interest you, because there is bi-cultural influence in my work. Ideas from the East and West coexist in my subconscious. In Japan, we have a culture rooted in several religious teachings. Shintoism, an indigenous religion, Buddhism, imported from China, and Christianity, imported from the west, coexist in peace. Interesting phenomena is that they co-exist in the lifestyle of a typical Japanese family. For example, a Shinto shrine placed above the kitchen space, a Buddhist altar in the tatami room to worship the ancestors, and sending children to Christian schools occur simultaneously in a household. We can conclude that the Japanese possesses dual personalities, rational and spiritual, seemingly a contradictory combination. I would like to illustrate this aspect of our culture from my works as a bi-cultural architect.

9.3.1 Villa Rosa

This project is a summer-house located in Karuizawa, a popular highland resort in Nagano Prefecture, about an hour from Tokyo by the Express train. The building is an 8.5 meters cube (Fig. 3), situated on a small lot surrounded by pine trees. What I sought to achieve in the design was for the building to function as a transition space for the owners arriving from the intense urban living to the environment of serenity. The sequence of movement from entering the property, ascending the ramp, entering the house, descending the steps to the living room would allow the owner to step over the threshold to the natural surroundings. In this white box, the users are cleansed of the accumulated negative urban experiences and "reborn" each time they visit to the house.

Architecturally, the cube is a weak volume. It is set up to express the process of "decomposition" of a manmade geometric volume and the building volume is expressed metaphorically to be disassembled back to nature. On the other hand, the users' spirit is rejuvenated in the natural setting. It can be said that there is an element of rationality and spirituality co-existing in this project. Moreover, the white color of the house blends in with the winter snow and reflects the summer sun. It follows a rational thinking, but the result offers the users a spiritual experience.

This building is within the bounds of the notion of contemporary space. It has a physically defined boundary. Walls

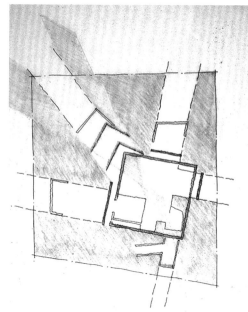

图3 采用白色立方体造型的夏季别墅，有机协调人工空间与自然空间

（资料来源：George Kunihiro）

Fig.3 Villa Rosa in the form of cube mediating the manmade environment and the surrounding nature

(Source: George Kunihiro)

图4 僧舍加建：屋顶形式的隐喻表达
（资料来源：George Kunihiro）

Fig.4 Addition of a temple: metaphorical expression through the form of roof
(Source: George Kunihiro)

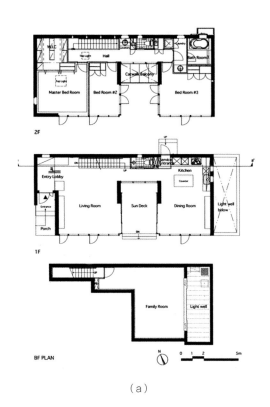

(a)

物质边界。墙体和窗户限定了室内与室外，特别是混凝土墙在限定空间的同时，给人以"强壮"之感。在室内，严格界定的内部空间迫使人们向外移动，走进室外的自然景观；另一方面，6米高的窗户又使自然进入室内，这就是借景的概念，景观设计上常用的技巧，利用花园中的树木或树丛形成景框，把远处的景致纳入其中，从而使花园的视觉和空间组合更加完整。坐落于树丛中的立方体建筑丰富了使用者的精神感受，使他们在轻井泽的短暂停留中实现了与自然的结合。

设计这个项目时正是后现代思想盛行的时期，我尝试通过这个建筑设计探索一条现代主义的地域化路径。

9.3.2 升起的荷花

这个项目是一座寺庙僧舍的加建，位于与东京都市区相邻的千叶县。寺庙属于中等规模的地方宗教设施，主要服务于佛教会众和当地居民。加建部分包括了僧侣三个孩子的房间，以及僧侣接待访客的非正式客厅。场地测绘时发现，场地上每栋现有建筑的屋顶形状各不相同，可能源于不同的建造年代。受此启发，我在设计中将屋顶形态作为重点。佛教视莲花为命运、纯洁和忠诚的象征，因此整栋建筑被设计成由两面互成角度的墙体和一个起伏的屋顶共同组成的莲花盛开的形状（图4）。建筑的两面墙体各偏离垂直轴线9°，相互交叉成楔形，表达生于地面的寓意；房顶采用了曲面形式，与相邻的幼儿园建筑的屋顶相呼应，孩子们每天早晨都会收到校长房间的建筑立面所传达的问候。窗户的弧形曲面表达了室内外空间的流动，特别是室内空间的外向拓展。建筑平面是简单的四方网格，充满理性，而建筑表达则富有象征性和精神性。

and fenestrations define what is interior and what is exterior. It is a "strong" building where concrete walls define the space. Defined nature of the interior space encourages one to move outward. From within, one feels the urge to explore out into the natural landscape. Simultaneously, nature seems to come into the interior through a six-meter high fenestration. This is the notion of "borrowed scenery", a landscape technique whereby far view is framed by the trees and bushes of the garden landscape to complete the visual and spatial composition of the garden. The house sitting amongst the trees is a box to process the spirit of the users to have them unite with nature for their short stay in Karuizawa.

The house was designed during the period of Postmodernism. With this house, I attempted to make a building exploring the regional alternative to Modernism.

9.3.2 Lotus Rising

This house is an addition to a monk's quarter in a Buddhist Temple located in Chiba Prefecture, adjacent to Metropolitan Tokyo. This local temple is a medium-sized religious facility serving the congregation and the local community. The program called for rooms for three children and informal reception room for the monk to receive the guests to the temple. When surveying the existing buildings on the temple precinct, we identified that the shape of the roofs were unique to each building, perhaps due to the varying periods of construction. From this observation, our design address the roof-scape and how the building was conceived. Lotus in Buddhism is a symbol of fortune, purification and faithfulness. Concept of the house is the blooming lotus flower shaped by two angled walls and an undulating roof (Fig. 4). The two walls of the building volume form a wedge, nine degrees off the vertical axis, expressing growth out of the ground. Shape of the roof comprised of curved surfaces, relating to the roof of the adjacent kindergarten building on the property. Each morning, the children are greeted by the expression of the façade of the headmaster. Bay windows have curved surfaces to express a flow of space between the exterior and the interior. From inside, the space of the room expands outward. The building plan is simple four-square grid and rational, but the building expression is metaphorical and spiritual.

9.3.3 Urban Horizon

The next house is a study in "non-existence". Urban landscape of Tokyo is cluttered, heterogeneous and lacks visual coherence, with so many shapes, colors and reference measurements. When I received this commission to design a residence adjacent to a farming lot in one of the 23 wards of the Metropolitan Tokyo, I decided to contribute to the urban landscape by dramatically reducing the visual impact of the building. The house is for a family of five with three pre-teenage to teenage children. Since the lot fronts the open field and faces south, I oriented the building towards this field (Fig. 5).

In the floor plan, all habitable rooms face south towards the field. The featured element of the design was a large freestanding movable louver wall placed 1.5 meters from the south facade of the building. Unfortunately, due to the high cost of construction, the owner decided not to build this louver wall. It was unfortunate, because the louver wall would have been functional and dramatic with contribution to green environment. With my failure to convince the owner to build the louver wall, I felt responsible as an architect. After completion of the house, I decided to turn my direction to architecture activism and stepped away from my practice for an indefinite period of time. It was a time of reflection. I spent my energy contributing to the research on the conservation and revitalization of modern Asian heritage. I involved myself in organizing numerous workshops in Southeast Asia, China and Turkey.

9.3.4 2@5

Five years had passed and one day, I got a telephone call from an American Expat living in Tokyo who inquired regarding my interest in participating in a small design competition for a house he was planning to build in Tokyo. I participated in this competition and won the contract to design the project. Compared to China, almost every aspect of urban life is compact in Japan, including urban lots and streets. This reality becomes critical and the architects must have the sense to design in tight spaces and still create spacious feeling and privacy.

The lot for "2@5" is an urban lot surrounded by houses built on three sides. Only the street frontage frees the building from physical proximity to other buildings. The final design ended

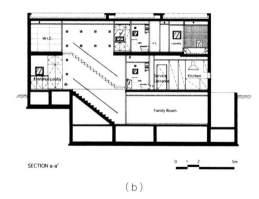

(b)

(c)

Fig.5 A housing design with respect to the urban landscape of Tokyo
(Source: George Kunihiro)

围，只有朝向街道的一面没有紧邻其他建筑。整栋建筑高四层，建筑面积200平方米，设计理念是"向上的刺"，以接近这栋建筑唯一可以获取的自然要素——天空。天空作为都市中的自然，使这栋建筑表达出向上的精神寓意。建筑室内的私密性也得到有效保障。从地下一层的书房和办公室，到一层的客房和二层的客厅，再到四层的主卧，整个空间被竖向交通紧密联系成一体；楼梯间最终通往屋顶平台，完成通往都市自然的旅程。这栋建筑的特别之处在于顶层的主卫，以及阳台上被木质百叶幕墙环绕的小花园；它并未遵循日本传统住宅把卫生间放在较低楼层的习惯做法，而是使其更接近天空，创造出漂浮在空中的都市自然（图6）。

9.3.5 镰仓的周末住宅

这个项目是一对美国和瑞典侨民的周末住宅的扩建，位于面向太平洋的历史城市镰仓。这个四口之家生活在亚洲十年有余，他们喜欢日本文化和亚洲生活方式，但要求新住宅必须拥有充足的自然光线。这个要求主要来自主人的妻子，作为瑞典人，她在文化上钟爱自然光线。他们第一次到事务所面谈时，明确要求设计一个玻璃盒子，以确保有充足的阳光进入室内。他们此前约谈的建筑师坚持认为住宅不应该是一个玻璃盒子，所以他们当时向我问了同样的问题："你可以为我们设计一个玻璃盒子吗？"我说"当然可以！"虽然心里知道，这可能是另一个不可持续的建筑。这一次，我把玻璃盒子包在一个百叶箱子里，这样既可保证足够的私密性，又可获得充足的自然光线。

我们把一半的基地用于建造三层高的住宅，另一半用于建造一个小型的日式花园和一个练习用游泳池。客厅和

up with a total of four stories and 200 square meters. Concept of the house was "Upward Thrust" to reach the sky, the only "nature" that this building can capture. Urban nature, the sky, makes this building vertical in spirit, while our design sought to guarantee privacy in the interior. From the study/office on the basement, to the guest bedrooms on the ground floor, the living room on the second level, then to the master bedroom on the fourth, the spaces are connected helically by the vertical circulation. The stairway finally reaches the roof terrace to complete the journey to the urban nature. Special feature of this house is the master bathroom on the top floor with mini-garden on the balcony protected by the wooden louver screen. Instead of putting the bathroom on the lower floor which is the customary location of the Japanese house, this building placed the bathroom closer to the sky, with a bit of the urban nature floating in mid-air (Fig. 6).

9.3.5 Weekend House in Kamakura MM

This project is a weekend house for an American and Swedish Expat in Kamakura, a historic city facing the Pacific Ocean, The family of four had lived in Asia for over a decade and decided to extend their residency in Japan. They are fond of Japanese culture and the Asian lifestyle, but asked for a house with abundant quantity of natural light. The request came from the wife who was from Sweden, a culture with true appreciation for natural light. When they first approached to our office, the owners asked us to design a glass box, with plenty of natural light entering the interior. The architect whom they interviewed to design their home before coming to our office insisted that the house should not be a glass box. They asked us the same question. "Can you design a glass box?" I said "yes!" knowing that I may be getting into another sustainable "unfriendly" building. Well, this time, I proposed to wrap a glass box with a louvered cage. This way, there is sufficient amount of privacy, at the same time, getting healthy quantity of natural light.

We designed a three-story house on half of the site, while creating a Japanese-style garden with a small exercise pool. Our design placed the living and dining rooms on the second floor with a comfortable balcony fronting the street. From here, one can view the garden and, with the setback, secure privacy from

图6 向往都市自然的住宅设计
（资料来源：George Kunihiro）

Fig.6 A house design towards urban nature
(Source: George Kunihiro)

餐厅被布局在建筑二层，通往面向街道的阳台；从这里既可观赏花园，建筑后退又保证了相对于街道的私密性，在玻璃滑门打开后还可以为整个家庭提供一个大型空间，就像日式住宅中的橼侧。起居室内，窗边的火炉成为家庭聚会的核心。三楼有两间卧室，属于私密空间，全部朝南，立面也被水平的木质百叶完全覆盖，以切断阳光直射，并保证卧室的私密性。经过百叶过滤的光线照进室内，将整个住宅变成一个模糊的空间。这栋建筑的设计基于西方理性，但是创造了富有东方意味的空间感受（图7）。

9.3.6 OGA盒子

这个项目是我想介绍的两个竞赛方案之一。OGA是指从木材上切下来的木片或碎木，木屑板就是由这些粗糙的副产品加工而成，但在此过程中产生的细小木屑往往被丢弃。我们决定做个实验，利用OGA木屑创造一种新的结构材料。我们的团队与一位结构工程师合作，设计了制作OGA木片和OGA木板的方案，进而利用这些结构材料设计了一个小型的、模块化、可移动、工业化生产的宿舍单元。这是一个边长5米的方形空间，内有起居空间和阁楼卧室，并配有厨房、淋浴和厕所，形成一个完整的OGA居住空间；其模数化的平面布局理性、极简，创造出类似于日本茶室的精神感受，即在最小的空间中创造出无限的宇宙空间感（图8）。

9.3.7 景福宫考古博物馆

这个项目是我想介绍的第二个竞赛方案，是事务所参加的韩国光州考古博物馆国际设计竞赛。韩国文化是与相邻的日本文化和中国文化相似的亚洲文化，历经数个世纪

the street. The sliding glass doors can be opened completely, like the "engawa" in the Japanese style house, forming a large space for the whole family. A fireplace by the window becomes a focal point for the gathering in the living room. Third floor is the private floor with two bedrooms, both facing south. This part of the facade is wrapped with horizontal wooden louvers to cut the direct solar gain, and to guarantee privacy on the bedroom floor. The house has an ambiguous space, where filtered light from the louvers illuminate the interior. The building design is based on rational thinking, but resulting spatial experience is rooted in the Eastern sense of place (Fig. 7).

9.3.6 OGA Box

OGA Box project is the first of the two competition proposals which I would like to present. "OGA" means wooden chips or particles produced from cutting of lumber at the mill. Particleboard is produced from coarse version of OGA, but the fine grade wooded dust is thrown away. We have decided to experiment and create structural material from the fine OGA dust. Working with a structural engineer, our team developed a scheme to manufacturer OGA lumber and OGA board. Utilizing this structural material, our team proposed a small modular, mobile, industrialized dormitory unit. The structure is a five-meter cube containing living space and a sleeping loft. The unit is equipped with kitchen, shower and toilet, making the OGA Box a complete living quarter. The modular plan is rational minimal. OGA Box has a similar spirit with a Japanese teahouse, a minimum space with universal spatial experience(Fig. 8).

9.3.7 Gyonngbok Archeological Museum

The second competition project is an international design competition our office participated for the design of an archeological museum in Gwangju, Republic of Korea. Korean culture is an Asian culture having many similarities with its neighbors, Japan and China. Over centuries, it has developed into its own unique culture. Working on a museum located in the natural setting, it dawned on me to look at the landscape and topography. I attempted to express the form and history of the landscape through the idea derived from "chima", the skirt of "hangbok", the traditional Korean female dress. The

图7 被木质百叶包裹的玻璃盒子，确保住宅的私密性与重组采光

（资料来源：George Kunihiro）

Fig.7 A glass box wrapped with wooden louver to ensure privacy and plenty of natural light

(Source: George Kunihiro)

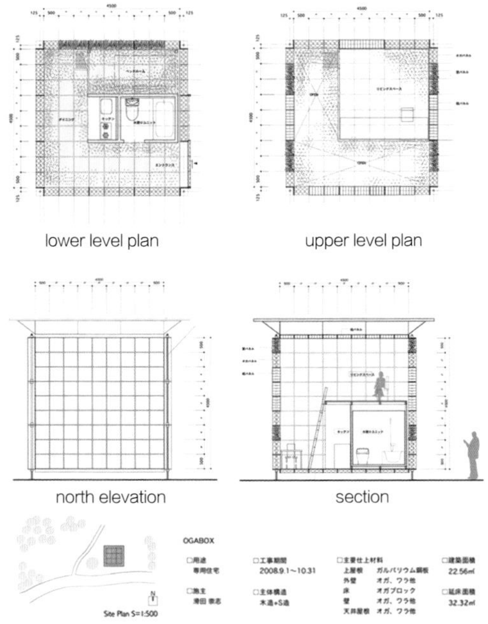

的演变形成自己独有的特点。博物馆所处的自然环境启发我审视周边的景观和地貌，力图通过源自chima的设计理念，表达景观的形式与历史。chima是韩国妇女的传统裙服，充分体现了褶皱的肌理和美感；褶皱的起伏加之织物的色彩，构成了整体结构的韵律。

我采用了chima的三维美感，将其应用于东西两侧都有山丘环绕的场地上；景观的褶皱与四季的颜色变化构成完美的背景，将建筑融入起伏的山谷之中。受到"挤压"的场地释放出压力波，与周围景观形成一种动态的关系。进入建筑，首先映入眼帘的是门厅中的"大洞"，展现出入口层下的考古挖掘现场；从大厅沿梯而下来到地下挖掘现场，可以真实感受到泥土和数千年时光的流逝；之后，参观者可以搭乘自动扶梯到达二层和三层参观展览，最终被引导到屋顶平台，鸟瞰整个场地的自然景观（图9）。

在利用传统"褶皱"理念造就建筑体验的过程中，我们的设计方案充分尊重了考古场地的永恒性特点；而在其他地方，时间依然不停流逝。建筑表达展示了时间的流逝。在自然中，人工的介入造就了博物馆的形态，它每年将带来成千上万人的参观流量，而场地的自然景观却静如止水。这个博物馆设计要表达的就是时间概念的融合，或是静止，或是动态，就像东方和西方之间的强烈对比。

9.4 结论

生活是一个连续的过程。经历了40年的建筑实践，我已习惯于以编年体的方式，讲述我在多种文化语境下的事业发展和作品创作。我们每个人都在各种条件下演进或变化。同时，生活又是一个不断积累的过程；我自己就曾

texture of "fold" and its beauty are best expressed in the Korean dress. Peaks and valleys in the "fold" give rhythm to the composition of the whole, especially with colorful fabric of these dresses.

I took this three-dimensional beauty of the "chima" and overlaid the idea to our site, which is defined by hills on east and west. Fold of the landscape, with changes in the seasonal colors, will be a perfect backdrop to our building hovering over the valley of the site. The site is "squeezed", which emit cyclical waves of pressure and release in the atmosphere, creating a dynamic relationship with the surrounding landscape. To experience this building, one enters and discovers the "hole" at the lobby. The excavation site is below the entrance level. From the lobby, one descends the escalator to the site of the excavation located at the basement level. Here the visitors can feel the earth and thousands of years history with the archeological excavation. Then, the visitors proceed up the escalator to the second and third levels to view the exhibition. At the end of the tour, the circulation leads to the roof terrace where the visitors are presented with a dramatic bird's eye view of the entire landscape (Fig. 9).

In translating this experience with a traditional idea of "fold" generating our building, our proposal respects the timeless quality of the archeological site. Simultaneously, time continues to move elsewhere. Our expression of architecture represents this passing of time. In nature, manmade intervention has been made in the form of a museum. It will bring thousands of people each year, while the landscape stands still. In this museum, the fusion of the static and dynamic notions of time are expressed as the contrasting of the expression of the East and the West.

9.4 Conclusions

Life is a continuum. Through my 40 years of practice, I feel comfortable to chronologically present my career and the works in the context of several cultures. We all evolve, or change, with each situation we encounter. But also, life is an accumulative evolution. In my case, I have traveled to over 65 countries and lived in two distinct cultures.

As the global community evolves into mobile community

图8 OGA盒子
（资料来源：George Kunihiro）

Fig.8 OGA Box
(Source: George Kunihiro)

游历超过65个国家，并在两种完全不同的文化中生活。

随着跨越国界的人员流动越来越多，全球社区将逐渐演变成动态社区，多文化阅历也将成为日常现象。我希望你们能够打开心扉尝试这种经历，同时又能保持自身传统，这样你的表达就不会消融在纯粹的世界语汇之中。

我期待，在未来，"一个亚洲"的愿景能够落地开花，枝繁叶茂！

with people going beyond national boarders, multi-cultural experiences will become an everyday phenomenon. I hope you can open yourself to such experience, while maintaining your own heritage, so that your expression does not dilute in mere universal vocabulary.

I look forward to the vision of "One Asia" blossoming in the coming generations!

图9 光州考古博物馆国际竞赛设计方案
（资料来源：George Kunihiro）

Fig.9 International design competition proposal for Gyonngbok Archeological Museum
(Source: George Kunihiro)